Face/On

Face/On

FACE TRANSPLANTS AND THE
ETHICS OF THE OTHER

Sharrona Pearl

The University of Chicago Press   CHICAGO & LONDON

The University of Chicago Press, Chicago 60637
The University of Chicago Press, Ltd., London
© 2017 by The University of Chicago
All rights reserved. No part of this book may be used or reproduced in any manner whatsoever without written permission, except in the case of brief quotations in critical articles and reviews. For more information, contact the University of Chicago Press, 1427 E. 60th St., Chicago, IL 60637.
Published 2017.
Printed in the United States of America

26 25 24 23 22 21 20 19 18 17   1 2 3 4 5

ISBN-13: 978-0-226-46122-9 (cloth)
ISBN-13: 978-0-226-46136-6 (paper)
ISBN-13: 978-0-226-46153-3 (e-book)
DOI: 10.7208/chicago/9780226461533.001.0001

Library of Congress Cataloging-in-Publication Data

Names: Pearl, Sharrona, author.
Title: Face/On : face transplants and the ethics of the other / Sharrona Pearl.
Description: Chicago ; London : The University of Chicago Press, 2017. | Includes bibliographical references and index.
Identifiers: LCCN 2016037671 | ISBN 9780226461229 (cloth : alk. paper) | ISBN 9780226461366 (pbk. : alk. paper) | ISBN 9780226461533 (e-book)
Subjects: LCSH: Face—Surgery. | Transplantation of organs, tissues, etc.—Social aspects.
Classification: LCC RD523 .P39 2017 | DDC 617.5/20592—dc23 LC record available at https://lccn.loc.gov/2016037671

♾ This paper meets the requirements of ANSI/NISO Z39.48-1992 (Permanence of Paper).

CONTENTS

1 · Introduction: Effacing 1
2 · Mind/Medicine 10
3 · Losing Face on Film 49
4 · Decoding the Face-Transplant Debates 88
5 · A Very Special Makeover: Face Transplants on Television 123
6 · Conclusion: Face Transplants and the Ethics of the Other 156

Acknowledgments 181
Appendix: Face Transplant Recipients 187
Notes 191
Bibliography 223
Index 237

CHAPTER 1

Introduction: Effacing

In David Lodge's wonderful academic satire *Changing Places*, we are introduced to the game Humiliation. The point of the game is for participants to reveal which classics of literature they have not read; the winner is the one whose gap is most humiliating. In the book, the winner (the loser) is a literature professor who has never read *Hamlet*. He wins the game, and he loses his job.[1] (The game has become rather popular in certain circles. You may wish to try it; I leave that time-suck to you.) Here's my own personal version of Humiliation: I've never seen John Woo's 1997 film *Face/Off* in one sitting. Appalling, I know. But in my defense (and leaving aside evaluations of the movie itself, which . . . I'll leave aside), the face transplant(s) that take place in the movie are fundamentally different from those I explore in this book in ways that are deeply significant. In the film, John Travolta's character switches faces with the terrorist portrayed by Nicolas Cage as part of a heroic quest to save lives. He doesn't want to do it, but he quells his misgivings and conducts this major personal sacrifice in service of the greater good. The face transplant at the heart of the film is done for the benefit of others despite the individual cost. It's a heroic masculine journey that will end when the world is saved. It's reversible. John Travolta will get his old face back. And while his experiences posing as a terrorist criminal mastermind may change him (and they do—spoiler alert), in the end, he gets his life back. He gets himself back.

Facial allotransplantations (FAT), commonly known as face transplants, are not reversible. While their effects may be temporary (faces, like all other transplanted material, can be rejected and lost), the old faces are, by definition, gone forever. The old selves attached to the old faces are perhaps gone forever. (But isn't that true of us all? Isn't every moment that passes a moment of our lives and our selves gone? And aren't we now different as a result?) And while face-transplant recipients may on some level undergo the surgery for others, to shield the world from their faceless condition, they are, in the end, doing it for themselves, for their own ease of life and ability to navigate the world. They may do it for their relationships with others, for their ability to develop relationships with others or deepen ones that had been compromised by facelessness and the trauma that led to it, but the surgery is ultimately not a heroic project, not in the save-the-world kind of way. Even though, as we'll see, it is sometimes painted as one. But its personal nature is often presented by bioethicists and journalists as cosmetic, and maybe even selfish, and maybe even cheating. The face transplant, in these terms, is an indulgence rather than a medical necessity, and we'll explore how these are two very different categories of intervention. Except, as I show, when they are not.

These two distinctions, the nonreversibility and the highly personal nature of the face transplant, matter deeply in the way that the surgery and its recipients are understood, presented, discussed, and debated. At stake here are the reasons that people get face transplants and the risks—both medical and psychological—that recipients undertake with the surgery. At stake here is the way that a new face may bring with it a new identity. At stake here is the very meaning of faces themselves. And the meaning of facelessness. That matters too.

This is not (simply, or even at all) a book about the historical stakes for the face. That topic is too, too big. People have thought about it, and written about it, well and in depth. Though I bring some of these theorists of the face to bear at the end of this book (Lévinas, first and foremost, and Deleuze and Guattari, Goffman, Sennett, and Butler), I don't wish to reproduce their work.[2] Also I can't; the long history of the face surpasses the modesty of my intentions here. Well, sort of. This is a book about the stakes for changing the face, and the changing stakes for the face. This begs the question of the meaning of the face, and I think about it with the help of these scholars, and carefully offer my own speculative theories. But rather than looking backward, I push forward, using the face and the most dramatic manifestation of its manipulations to date, the face-transplant surgery, as a way to think about the self, and its always-public representation, as always changing. As always becoming.

I approach the face transplant through the framework of the index, the indexical relationship between the face and character. When that index is vexed, when the context for the correlation has changed, what do our reactions tell us about the meaning (for us) of the face, and the flexibility of its symbolic features? And what about when that index is vexed in the most extreme way, by substituting not just another face, but the face of another? What does that tell us about both the face and the condition of facelessness? That's my question. My answers are in the book that follows, but I'll give you some hints: the index of the face has always been unreliable. The face and the self are always changing and have always been always changing. The face transplant brings that unreliability to the fore, as our reliance on the index itself is being undermined. As it should be. We should encounter the faceless just as we encounter the face. Faces are not the index to humanity, no more so than facelessness. We owe the faceless the face-to-face encounter as well, and the ethical obligations incurred therein. We don't always live up to these obligations to the faceless; we don't always extend to them our acknowledgment of their basic humanity. That might be changing; it certainly ought to.

So how should we (re)conceptualize the face while still taking very seriously the importance of embodied experience? That's what, in part, this book tries to figure out. Like all processes, it's a journey. Writing this book, I've gone on a journey. It started in the nineteenth century, in Britain, with a graduate-school project on faces and their relationship to character. It became a book, and I became an academic.[3] And that book thinks about the face as an index, ultimately arguing that our reactions to the faces of others tell us more about ourselves than the people around us. It also thinks about how people self-consciously fashioned their faces and appearance to play to the index. Maybe to manipulate it.

I took those ideas and assumptions to this book, also about faces. (It turns out—and I truly didn't know this before—that I'm a little obsessed with faces. With the face-to-face.) But what I knew about faces in the nineteenth century turns out to be wrong in the twenty-first. It turns out that while some important things stay the same—bodies matter, faces matter, our relationships to our own faces and to the faces of others matter so very much—other things have changed. And at the end of this book, I outline not only how the face and its status have changed but also how my own thinking around the face and the interlocutors of the face has changed. As I hope you'll see, this is a very personal project in addition to an academic and professional one. It's a project of fighting for the voices and the humanity of the faceless and disfigured, the visually different and disabled, and thinking about why they need fighting for. Thinking about why we have denied the humanity of the faceless or, rather, why we imagine humanity to be vested in the face, and what happens when that assumption is shown to be false. In the conception of the face as index, the faceless are left out. They are, in a way, left for dead.

The faceless are not dead. But at the same time, their new faces bring with them new possibilities, new ways of imaging what bodily change (which we all undergo, though not to this extent) might mean.

This is a project of reframing how we think about minds in relation to bodies, and how we think about interventions on the body with respect to the mind. I hope to unsettle the division between mind and body by showing that treatments to the body treat the mind, or can; minds are deeply individual things. And (like all acts of writing, maybe) this is a project of passion and a project of love. This book about faces is me at my most intimate and me at my most public. I defer some of my own voice throughout the book in favor of my evidence and interlocutors, but at the end I chart what I've learned about the change in faces, and how my own thinking has changed.

The face was always going to be a deeply personal project. What can be more intimate, more inviting, more vulnerable, than sharing our faces? But at the same time, what could be more performative, more produced and manufactured, more public, than the faces we present to the world?

There is no contradiction here. The face is relational. That's one of the major themes at the heart of this book: the face is always public. Manipulations to the face are an act of publicity, one that we think is final. It isn't. We think of the face as fixed. It isn't. Not only because we can manipulate it, operate on it, transplant it, without necessarily rewriting history or transplanting ourselves. But also because we are all always in the act of becoming. As are our faces: they too are always changing. They too have multiple manifestations.

There are many different kinds of faces. But that doesn't mean we should lose sight of the centrality of the actual face, which lies at the heart of my book. Metaphor is important, but so is flesh and blood. In what follows, I use the face-transplant surgery as a way to explore the supposed indexical nature of the face by examining when the index is vexed. Face transplants are a particular case of facial manipulation because they not only change the face, they change it into the face of another. A random other. The story of the face transplant is not the story of changing the exterior to (finally) match the interior, a narrative that corrects faults in the index rather than challenging the index itself. Rather, it's the story of fundamental disruption. At least, it could be. It could be the story of radical hybridity, a new way of thinking about the body and our relationship to it, a new version of the power and limit of self-fashioning and representational judgment. It could be that story, but it isn't: instead, as I show, it's the story of neutralization, of making even this most disruptive of interventions into yet another makeover big reveal conforming to standard gender narratives and bodily norms.

But the face transplant isn't—exactly—cosmetic surgery. It also isn't—exactly—a whole-organ transplant. It's maybe a little of both, and versions of the story have been told to support both formulations, but I think that in the resistance of the surgery to categorization lies an opportunity. We can now trace the shape of a third thing that is a new way of conceptualizing the body and its interventions. It's a way to unsettle the distinction made between bodily interventions and psychological changes. It's a chance to reimagine what the face means without eradicating or ignoring its representational power, its links to identity, and its archiving of history. But we're not there yet, as this book shows. And much of what follows is the chronicle of us not being there yet, as I trace narratives of the face transplant as its resistance to categorization is negotiated and neutralized. I track this potentially radical and certainly pathbreaking surgery from its initial representation as a dangerous and shallow cosmetic procedure through its normalization as a technology of self-actualization and personal growth. I look at the various pieces that contribute to its (sort of) changing public presentation, including early fears of dramatic self-transformation and monstrosity, journalistic framing practices, rhetorics of medical fixing, military involvement and rehabilitation, and makeover narratives. I then imagine, alongside other dissenting voices in this history, another way for face-transplant surgery to be understood.

I'm playful with my sources, moving from medical literature and surgical history to bioethical debates, journalistic reports, reality- and news-television programming, and highly theoretical treatises on the meaning of the face. I dive into cultural manifestations of facial manipulation and substitution, looking in particular at transplant films and the fears they both represent and produce. Undergirding it all is an engagement with the literatures and theories of difference and disfigurement, around disability, around gender, around monstrosity. I think about hybridity and the possibilities and dangers therein.

Face transplants seem to be the realization of medicalized mixing and melding, unsettling stabilities and creating liminal and collective spaces, bodies, stories, and institutions. The lack of fixed points in what is a highly detailed and tremendously complicated surgical intervention makes us very uncomfortable. In a story about identity, we can't quite tell who the main players are because they themselves are not discrete. Or, to put it another way, it is not about the difference between the cyborg and the monster but the difference between the cyborg and the goddess.[4] At the moment, there isn't a real difference between these two transformed productions. But, as the face transplant and other examples that I chronicle show, there could be.

The before-and-after, the big reveal, is a well-known moment in all makeover contexts, televisual and otherwise. In the case of facial-allograft surgery, the big reveal is ongoing with the possibility that the new face may disappear ever-present. (Now that would be a reality moment.) The trauma is always on display; the scars are always visible. What is so shocking about these particular transformations is the normalization of that trauma. These deeply scarred people start to look like everyone else. The transformations by which they earn the right to their new selves lie in precisely the banalization of their pain; when they say they would not go back to the time before their face loss, when they say they have changed, improved, become too different for that, only then can their facelessness disappear. It's a highly conservative narrative, and one that serves to neutralize the radical possibilities of this kind of hybridity. It makes this kind of transformation, like the reality narratives it echoes, almost inevitable. With the right kind of personal labor.

Onscreen for everyone to see.

No other surgery has been archived so richly and in such a wide variety of highly sophisticated media. This is partly because it is so recent, and its documentarians have access to the most advanced technology. It is also partly because of the nature of the procedure itself; visually awe inspiring, the surgery is at its core about giving someone a new appearance and, in so doing, a new or often renewed life. But while many makeovers are ostensibly transformative and may even change people into a new version of themselves, most do not go so far as to turn their subjects into actual other already-existing people. We are left with the question of what the face transplant actually transplants, and who the new person actually is. So much of our identity seems to revolve around our faces; does the face of another bring with it the identity of another?

I use this question as a way to think about how the self has always been in process, always changing, and always mediated. The self has always been a set of choices, and this book discusses what we don't and can't always choose, namely the reactions of others and how they relate to us, and what we can choose, namely how we imagine we ought to relate to others. And that imaginry, thanks partly to the face transplant and its associated technology, and thanks partly to the multiple representations of self in the digital environment, is changing.

More concretely, in chapter 2, "Mind/Medicine," I turn to the history of cosmetic surgery and its intersections with transplants to start the story, showing how FAT is both and neither and something else all together. Through a deep dive into the literature of cosmetic surgery, I map the discursive terrain on which face transplants have been charted, thinking about the rhetoric of facial manipulation in the context of FAT. If the lack of a face is a debilitating medical condition for which a transplant is the cure, what is the difference between that and a more purely cosmetic intervention—is it a difference of degree or kind? The narrative differentiating the former from the latter, as we see in this chapter, is one of risk: the risk of the operation and especially of the lifetime on immunosuppressants while living with a new face. But maybe risk is just an excuse that obscures the true source of our objections. I focus on fears of identity transfer and cellular memory, tracking the literary and cultural manifestations of this phenomenon, and think about how these impact conceptions of transplant surgeries more broadly.

Chapter 3, entitled "Losing Face on Film," moves to cinematic representations of face transplants and the fears and concerns they reveal in three films from the 1960s, *The Face of Another*, *Eyes without a Face*, and *Seconds*. These films, all shot post–World War II, investigate the varying cultural considerations around losing face in Japan, France, and the United States, respectively. I show that in each film acts of radical self-transformation are punished, even if they represent an escape from the liminal state of facelessness, a state between life and death. And they do; face transplants are an escape from being the living dead. I analyze common themes in all the films, thinking about the role of medicine and the evil doctor, the dismantling of the family, the stakes for self-transformation, the adherence to gender roles, and the liminal space of life without a face.

In chapter 4, "Decoding the Face-Transplant Debates," I go to the ground, exploring the popular, medical, and bioethical debates following the first partial face transplant of Isabelle Dinoire in 2005. Here I concentrate on the technical aspects of the surgery, as well as the way that it was framed by surgeons and bioethicists during early attempts to get ethical approval for the FAT. These were careful rhetorical negotiations, rendered almost null when the French team announced the completion of the Dinoire surgery. I show that there was space in the initial journalistic coverage of the surgery to imagine it as something other than strictly cosmetic and strictly self-serving, considering and taking seriously the therapeutic possibilities of facial interventions. I then look at how and why that space constricted as Isabelle Dinoire and her doctors were put on journalistic trial for attempting to win what newspapers around the world dubbed "the face race" (a trial that they lost).

Chapter 5, "A Very Special Makeover: Face Transplants on Television," looks at the continuation of the face transplant through its mediatization. I focus in particular on television representations of subsequent face-transplant recipients in the United States, reflecting on the highly conservative, deeply normalizing stories these representations tell. I map these television representations onto both the medical-documentary and makeover formats. This analysis shows the rhetoric of the big reveal of the conforming external self, earned by trauma and the work of self-improvement, which renders the face transplant yet another version of making the exterior—the parts of the body we see—map onto the interior—the hidden and black-boxed and impenetrable self. There is a specific national context to this study, as the United States is the only country for which recipients of the surgery needed a private source of funding. There's also a gendered story: women make better (makeover) television, a significant advantage in raising needed funds and awareness for the procedure, possibly accounting for the greater number of female recipients in the United States than the rest of the world. I consider the role that television portrayals play in making the surgery palatable and acceptable, a trajectory heightened by military involvement in, and funding of, the surgeries and their associated research. I argue that military rhetoric frames the surgery as the solution to the social problem of the injured vet. This chapter shows how this potentially radical intervention is made deeply normal and, in its way, unproblematic. Except that, in the end, the index between the face and the self is fundamentally vexed. Except it turns out that it always has been vexed, and we need to rethink what we imagine faces can tell us.

That's what I think about in the final, concluding chapter, "Face Transplants and the Ethics of the Other." There, I look at public performances of facial and bodily manipulation by Orlan, Genesis P-Orridge, Michael Jackson, and Joan Rivers and how these bodily canvases open up arenas for new ways of thinking about the body, the face, and the self. I track how, through the course of the research for this book, my own thinking about the meaning of the face and its indexicality has changed. I think this change through the various theorists of the face whose ideas intersect with my own, idiosyncratic, deeply personal interests around ethics, feminism, science studies, and the media. I argue that it's time to come up with a new ethics of the face, a new conception of what faces mean and what we mean when we look at faces.

To turn to another literary source, I have asked—often, when writing this book and in other research: to show or not to show (images)?[5] My argument stands without visual representation of the face-transplant recipients in all the phases of their faciality; to reproduce these already widely available images plays further into the centralizing of appearance and the stigmatizing of difference. To show images is to look at images. To gawk at images. To stare at images. To forget about the people represented therein as they become only their (lack of) faces. At the same time, the visual, and the stakes for the visual, are central to the very stakes on which this book rests. The absence of images may be a kind of distraction that begs for a kind of abstraction that in its own way negates the very real and very specific nature of the stories that I tell. And to not show images validates a kind of distancing from these people, allowing others not to look. Not to stare. Not to see the people represented therein.

So, to show or not to show?

Yes. I both show and do not show, offering images of the recipients prior to their injuries, following the damage to their faces, and post-transplant in an appendix to this book. If these photographs are integral to your understanding of what follows, I give you the tools to complete the picture. And if you need even more than what I have reproduced, Google has given you even more tools and even more choices.

A numbers check: only thirty-one or so face transplants have actually been performed, which isn't really a whole lot given the possibilities I imagine for this procedure. But, as I show, the face transplant speaks more broadly to questions of the self, the relationship between mind and body, cultural negotiations of self-presentation and manipulation, and the multiple arenas engaged in these questions. These arenas range from audience studies that manipulate faces in advertisements and other media to elicit behavior from viewers, to design of digital avatars and other identities online. And these arenas, disparate as they are, have something important in common: they all engage with the problem of distance. The problem of relationality at a distance. I conclude this book by turning away from looking and from the relationships and publicity engendered by looking. I turn instead to touching. To practices of touch. And I think about what this sensory relationship, what this particular mode of communication, has to say to the face transplant and the manipulation of faces. I think about what touch makes of the index between face and self, and what new kinds of indices emerge from literally and physically reaching across the distance.

But it is not to touch that I turn now, not at first. Let us begin with sight, with looking, with reading. Books (like faces and people) are always in the act of becoming. So, please, read this book. And, in so doing, change it.

CHAPTER 2

Mind/Medicine

INTRODUCTION

In 1994, nine-year-old Sandeep Kaur was working in the fields in Northern India when her hair got caught in the thresher, ripping off her face, scalp, and one ear. Her mother quickly put the two pieces of her face in a plastic bag and raced her to the nearest hospital, a three and a half hour moped ride away. Kaur underwent a nine-hour surgery, during which microsurgeon Abraham Thomas replanted her face in the world's first facial reattachment. To date, Kaur, who is studying to be a nurse, bears facial scars and lacks some mobility, but remains recognizably herself and has many facial functions.[1] It was Kaur's own face that was reattached; unlike face-transplant recipients, she didn't need to be concerned with blood-type matching, a lifelong immunosuppressant regime that would significantly compromise her health, and the psychological impact of bearing the face of another. The bioethicists, therefore, did not care. The medical community, outside those who would later chase the facial-allograft dream, hardly cared.[2] The press and the popular imagination certainly did not care. Kaur's unprecedented surgery caused hardly a ripple, let alone the tsunami evoked by Isabelle Dinoire's 2005 partial facial transplant. Getting one's own face back didn't make one a medical or media spectacle. It didn't make one a monstrous hybrid of self and other, a locus of ethical debate, and a testing ground for the seat of self-identity. Beyond the attention garnered by facial scarring, it barely made one interesting.

I start with the story of Sandeep Kaur not for what it initiated in the form of later facial transplants but for what it did not, namely a storm of controversy and bioethical debate. The lack of interest around Kaur's surgery did not stem from a lack of medical innovation—the procedure was the first of its kind to be performed and required skillful and painstaking effort as well as innovative techniques.[3] Kaur's geographical location, away from major medical research centers in the early 1990s, kept media attention relatively low, but that is only part of the story. People did not care, or did not care all that much, because the surgery itself, as well as the person who emerged from it, transgressed no boundaries around the stability of the self, the nature of identity, and the relationship between character and the face. Kaur's reattachment surgery was one that attempted to recapture the person she had always been, rather than to make her into new one.[4] She wasn't attempting radical self-transformation. She wasn't cheating the dictates of nature but restoring them. And while her face was now more scarred and less mobile, and while this certainly would have implications for her future, she was still much better off than had her face not been reattached at all. The question of Kaur's future was particularly salient for one so young, whose age also protected her somewhat from the narrative of ennoblement often tied to disfigurement and suffering.[5] And while this surgery may not have been strictly lifesaving, the psychological and medical risks of a reattachment procedure were far fewer than those presented by a facial transplant. Which is to say: the risk-benefit analysis (were one done) would have weighed heavily toward benefit, despite the considerable risks of an untried procedure. This was not an ambiguous case. It wasn't unclear whether or not the surgery was worth it or if, from a health, psychological, or identity standpoint, the procedure would make things worse for Kaur. Not that Kaur's doctor had time to consider these questions—to have any chance of success, he had to dive right in.[6]

And he did. Of course he did—he had the opportunity to give a little girl back her face. To give a little girl back her life . . . if not quite as she knew it before, as least in a way that wouldn't have to change how she knew herself, and how others knew her. It matters, too, that she was a little girl; in cases of extreme damage or defect, children have a lot more leeway to manipulate themselves.[7] In cases of minor difference, they—and their parents, of course—have far less.[8] And, yes, she was in India, far away from Western centers of medical development and debate. That also matters, but not as much as we might think; the next case of facial reattachment that we know of involved an adult woman in Australia. Her operation too was largely ignored.

In 1997 an Australian woman's scalp was torn from her head by a milking machine.[9] In its report of the event, the AP article quoted the lead surgeon, Dr. Wayne Morrison, who pronounced the surgery a success while acknowledging that the patient might need additional surgery to deal with scarring. Notably, Morrison's assessment of success was based on the fact that "she will still be identifiable and have her own personality.... The essential characteristics of her face will be there."[10] According to Morrison, the surgery enabled the patient not only to remain identifiable as herself but also to maintain her personality, implying that the former facilitated the latter. Doubtless the challenges of severe disfigurement and pain could cause fluctuations in a patient's personality, but Morrison was not here referring to the patient's avoidance of such challenges. Rather, he was arguing that the patient's maintenance of her own face allowed her to maintain her own personality; were she to look dramatically different, were she to no longer be identifiable as herself, she would no longer, at the level of character (whatever that means), be herself.

While we must take these postsurgical remarks to the press with a grain of salt, Morrison is not alone in linking one's face to one's deeply personal character and sense of self. Or, by extension, the loss of the face to the loss of character. Or, by extension, the adoption of the new face as the adoption of new character. This is a narrative binary that we will see repeated between interventions that restore a person to the self she once was, or the self she ought to always already have been, and those procedures that turn someone into an entirely new person, someone he otherwise never would have become. It's also an incredibly (if unsurprisingly) potent demonstration of the power of personhood that we vest and rest in the face. Which is part of the reason why manipulations of the face can be so fraught: the face itself is so much more than features. The face, as we imagine it, is an index. There has to be good reason to vex it. Part of what I do in this chapter is explore what counts as a good, or good enough reason, and what doesn't. And why and how that has changed, and changed again.

That's the general focus. More specifically, I map the discursive terrain on which the face transplant is discussed and negotiated, analyzing the ambiguity that surrounds its status as a medical and therapeutic intervention. More than that, the face transplant resists categorizations and must, I show, be treated as its very own, very specific kind of thing. It is both organ transplant and cosmetic surgery, and at the same time it is neither, even as it shares characteristics with both. It both makes people better, improving their appearance and dealing with associated psychological challenges, and makes them worse, significantly weakening their immune systems and long-term physical health through antirejection medication. It is necessary and maybe even lifesaving for some and not at all for others, and the difference lies not in the subjects' biology, and not even in their psychology, but in the interactions of these with the world. Some need to minimize facial damage more than others. Unlike with (a straightforward medical approach to) whole-organ transplants, the therapeutic implications for the face transplant are highly individual, and rest on factors that cannot be consistently measured across cases.

A new face is a dramatic change that can bring about a shift in identity; the face of someone else brings with it not only change but also the identity associated with someone else. So here, too, it is both like and utterly unlike cosmetic surgery and whole-organ transplants. There are also factors not entirely dependent on the patient alone but also on society's reaction to her at large. These are a lot of variables, and a lot of unknowns. And while medicine, and transplant surgery, has certainly trafficked in the experimental and unknown, the boundaries surrounding acceptable risk are significantly lower for supposedly nonlifesaving interventions.[11] So the procedure's resistance to categorization had and has very real stakes in terms of how it is understood and what it is imagined to be able to do. And if it was (and is) allowed at all.

To locate these questions, I'll turn to cosmetic surgery and think about what it is and how it overlaps with facial transplants, ultimately showing the limits of this framework. I'll do the same for organ transplants, highlighting in both cases the stakes for identity and the role of the face as a privileged site for this investigation. Along the way, I'll consider the status of the potential recipient and the variant stakes for her refacing, using debates in disability-studies (itself a highly unstable and variable category) to insist on a phenomenological and non-Cartesian approach to the relationship between face and character and the role of one in impacting the other.[12] We'll see that underlying the concern about these medical interventions is a fundamental concern with the nature of identity as it relates to, on the one hand, integrating parts of others into the self and, on the other hand, deliberately altering the face as the visible index to one's internal nature. As part of this discussion, we'll consider cosmetic surgery and its relationship to beauty norms, psychological debates about body anxiety, and the traditional (and disputed) binary between necessary (read: reconstructive, physiologically mandated, or virtuous) and unnecessary (read: aesthetic, elective, or wrong) plastic surgery. We'll discuss the ways that this binary impacted understandings of the facial allograft, complicated by the special status of the face itself. Then we'll turn to whole-organ transplants, those that are, broadly speaking, medically necessary, and think about how face transplants might fit it. We'll tie this perspective to a discussion of the birth of bioethics and its relationship to organ transplants, asking, exactly what kind of (if any) medical therapy is offered by the face transplant? Or, more broadly, can changing the face and body cure the mind, and ought it to?

But *does* changing the face fundamentally change (whether curing it or not) the mind, the soul, the character of a person? And, to complicate the question further, does transplanting parts of another change the character to that of another? Does the question even matter? Well, that's easy enough to answer. Yes, it matters very much. I track our concern with these questions through just a small sample of the overwhelming number of cinematic treatments of bodily manipulation and the inevitable transfer and change of character that accompanies it. As part of the discussion in this chapter, I'll examine the fear of identity transfer or cellular memory (the notion that organ cells themselves retain memory that can be transferred to a new host), not in service of a medical analysis but as a way to consider our relationships to our own bodies and particularly our faces, and how deeply we feel our characters are tied to our somatic selves. And, how vulnerable we seem to feel that tie truly is, given our concern that it can be severed and adopted by another. I focus on narratives of identity transfer that emerged well before the face transplant was a technical possibility. The age of these examples is deliberate; it shows that we've been worrying about this for a long time.

And, as I argue here, we haven't figured it out yet. We can't make sense of the identity of the face-transplanted patient, who lives forever in this ambiguous space as recipient of an intervention that is both and neither cosmetic and/nor medically therapeutic. The melding of two humans, of what we imagine to be the melding of two human characters, is exacerbated by but not unique to the status of the face itself. The underlying need for a face transplant destabilizes our notion of what it is that medicine does and how therapeutic interventions themselves work. That's before the surgery. Then there's the after. We encounter this new kind of freak or monster, a person whose identity either has at best changed or at most is now an identity of more than one: a person either without a fixed psyche or with a radically changed one. This mutability transgresses our ideas about categories of people, and it also transgresses our ideas about categories of medicine. Maybe biology is as unpredictable as psychology. Maybe all medicine is dependent on the individual. Or maybe not.

So I suggest another way. Much like the face of the transplant recipient is neither the donated face nor the preinjury face but a third face, so too is the surgery a new kind of thing.[13] A third way. I follow the lead of others to explode the binary between virtuous and gratuitous self-manipulation. I reject the split between physiological and psychological intervention. Instead of locating face transplants somewhere in between cosmetic surgery and whole-organ transplants, some kind of happy and safe middle ground, I propose face transplants to represent an entirely new set of possibilities for thinking about bodies, faces, surgery, and their relationship to identity. And (hell, let's be ambitious) even reimagining an entirely new way of relating to others. An entirely new way of being human.[14]

THE STORIES OUR BODIES TELL US

The body itself is a special kind of communicative device that has been read as text or, as Carolyn Marvin and others have emphasized, must be understood as a form of experience and political and physical force.[15] Both, maybe. The statuses of the stories of the body are themselves ambiguous, highly mediated and manipulable as they are. As we refine and expand our ability to more seamlessly intervene on bodily surfaces and depths, our relationship to physical communication must be reevaluated. Even as we ask what kind of medicine the face transplant is, we must ask what kind of index the body is now, and what kind of index we have imagined it to be. The face and body as archive, as in Allan Sekula's formulation, underscores the careful collection and overarching organization that goes into the knowledge contained therein.[16] Medical symptoms and injuries, however, are often read as uniquely trustworthy historical and experiential narratives and indices. Colleen Farrell has surveyed the role of lesions in telling the story of Kaposi's sarcoma and, through embodied communication, the development of AIDS and its related history.[17] Sander Gilman has chronicled instances of reading deformity as recounting of history and the fault lines therein.[18] And our (read: Western, privileged) conceptualization of visual difference as deformity, and problematic deformity, rests on a particular set of medical, social, and policy structures with their own histories and roots, as Susan M. Schweik has demonstrated.[19] There are stakes to overprivileging disfigurements as transparent narratives; these two can be manipulated and have multiple meanings. The face is an ambiguous source. Manipulations of the face are ambiguous kinds of manipulations.

Bodies and faces tell us all kinds of stories, some of which are fabricated or disrupted. As the *LA Times* noted in 2005, "we're tossing out our own histories in favor of noses or cheekbones that may, perversely, reflect the values of our current society that may have nothing to do with our heritage, our families, or even sometimes our ethnicity. It's yet another triumph of the present over the past. Why look like Aunt Hilda when we can look like Paris Hilton?"[20] If we deliberately change our bodies, are we eradicating or manipulating history, or reliving and redefining it? This is one of the questions that Donna Haraway asks in "The Cyborg Manifesto," though for Haraway the recognition of imperfect communication is one of the affordances of the cyborg body.[21] She heralds the disruption of the historical narrative of the body as a great value of the cyborg, a composite and powerful being that embraces the vexing of the index and dispels myths of bodily limitation; other scholars, including Susan Bordo and Rosemarie Garland-Thomson, greet bodily and especially facial manipulation with rather more suspicion as attempts to eliminate difference and produce an idealized normal appearance.[22]

At the heart of the concerns around facial manipulation—through cosmetics, surgical intervention, or even dramatic changes in behavior—is a concern about character manipulation: if you change your face, do you change who you are? And, underlying this abstract philosophical concern is a more practical one: if you improve your appearance and, in so doing, appear to be better than you are, is that cheating? To put it another way, when faces are subject to change, how can others truly know who you are and, importantly, who you have been? As Sander Gilman has examined, the prevalence of syphilis in the fifteenth through the nineteenth centuries caused many to lose their noses due to the disease. Widespread noselessness both drove research and developments into nasal prosthesis and generated significant moral disapproval: lack of a nose indicated questionable behavior of the sort that caused the contraction of syphilis. Even those who lost their noses in other ways were painted with the brush of moral disapproval.[23] But for those who got effective nasal prostheses or surgeries, how was anyone to know of their behavioral depravity? Cosmetic interventions represented a disruption or even circumvention of bodily communication. Anxieties around cosmetic interventions and their disruption of communication have scripted historical objections to cosmetics, whose use was, as Annette Drew-Bear has shown, allied with corrupt morality and devilish motivation in the early modern era.[24] Kathy Peiss's book on the history of cosmetics has extended the discussion of the suspicion of cosmetic manipulation through the modern era.[25]

Participants seek to change appearances that do not accurately reflect their personality and character through surgery or other means. The narrative here, exemplified by the makeover show big reveal, is that appearance does reflect inner character; a faulty correlation can and ought to be realigned as part of the neoliberal construction and marketing of the self.[26] In these cases, such a correction does not constitute the kind of bad faith so historically derided but, rather, acts as a corrective for a problematic and ultimately highly limiting situation. It isn't, in other words, cheating. Gilman has raised the stakes on this question, charting the development of the distinction between so-called "good" or medically warranted plastic surgery and its opposite, the "bad" or gratuitous version that is depicted as nonmedical, nontherapeutic, and indeed dangerous in its own right as an unnecessary and risky intervention. He effectively problematizes this division, showing the institutional reasons for marginalizing plastic surgeons and arguing that aesthetic surgery can be thought of as (sometimes lifesaving and at the very least highly important) psychotherapy. In this work, Gilman charts the way different theorists have shown that improving one's body can improve one's mind, which can in turn extend (not just enhance quality of) life.[27] Reconstructive interventions that produce cosmetic effects, such as deviated-septum nose-job surgeries and breast-reconstructive surgery, highlight both the practical and intellectual difficulties of drawing sharp divisions. Cosmetic interventions do not stand in binary opposition to lifesaving surgeries; they are on a continuum of identity making, refining, and possibly correcting, as Diane Naugler's work on breast-reduction surgery puts in sharp relief. Her analysis argues that the division between cosmetic and reconstructive serves to further instantiate sexual and aesthetic norms, when in fact there is often very little meaningful division between the two.[28]

Face transplants present a staunch resistance to defining what kind of medicine they actually are, cosmetic or therapeutic. A number of scholars have located that same ambiguity in cosmetic surgeries themselves, seeking to trouble the very binary between interventions that are aesthetic and those that perform a healing function. This approach challenges the binaries between, let's say, virtuous (like Sandeep Kaur) and nonvirtuous facial interventions. Sander Gilman and Kathy Davis have long argued for a new understanding of the meaning and motivations behind cosmetic surgery, situating it squarely within the therapeutic framework.[29] Gilman shows that cosmetic interventions can have very real mental-health benefits; Davis explores the question of candidates' reasons, demonstrating that most recipients of plastic surgery are on a quest to correct perceived abnormalities rather than to become extraordinary. Recipients just want to be normal.

(The best version of normal, of course. There is an analogy here to psychopharmacological treatment that corrects for chemical imbalance and restores mental health—sometimes. And sometimes the non-ill take antidepressants to become better than well. But it's not all that different; it just redefines the locus of the perceived lack and posits a new definition of normal. The best version therein.[30])

According to Gilman's formulation, we can situate the face transplant as a kind of cosmetic surgery, which, its candidates might insist, in no way detracts from its therapeutic (and medical) necessity. And that's right, except in the ways that it is wrong. Medically, of course, the aftermath of the face transplant is entirely different: antirejection medication makes patients more vulnerable to illness.[31] And the meds themselves are carcinogenic. That's one difference. Sarah Kember and Joanna Zylinska makes similar moves in their research, positioning the face transplant as both continuous with and radically different from cosmetic surgery. It is continuous in its instantiation of self-identity and it is a departure in its destabilization thereof; the postoperative patient is not restored to a former, prefaceless identity but must establish a new one.[32] So they acknowledge the problem of categorization. But they don't solve it; they don't tell us what the face transplant *is*. They can't; we don't have the language for it yet.

Bernadette Wegenstein and Nora Ruck turn to the rhetoric of self-improvement, arguing that modern makeover culture is predicated on a perceived mismatch between external appearance and internal character; interventions are designed to align the incongruity and correct the error.[33] Katherine Sender's research on the makeover supports this approach; she shows that the genre negotiates the problem of making manifest the true and authentic inner self in a way that is visible and meaningful in the wider world.[34] Heather Laine Talley offers the most extreme version of this narrative, arguing that disfigurement is positioned as social death, a crisis sutured by what she calls "facial work," a category that includes elective aesthetic surgery, reconstructive work, and face transplants.[35] For Talley, facial work is not only therapeutic but also lifesaving. Talley's work makes explicit the new medical landscape to which these scholars gesture, arguing that facial work reframes the role of surgical interventions and, indeed, medicine itself.

All of these studies are vested in an understanding that manipulations of external appearance have very real and meaningful implications for individuals and their sense of self. While these scholars differ in the extent to which they believe these manipulations can be related to pathological states of mental health, their findings all argue for an expansive and expanded notion of the work that self-change can do. All these studies push back against the notion that cosmetic surgery is frivolous or self-indulgent, which rests on an assumption that these procedures are inherently or self-evidently problematic, a perspective at odds with neoliberal values of self-improvement and reconstruction.[36] This scholarship has tremendous implications for the face transplant. The surgery, and reactions to it, in many ways offer a test case of the posited polarization between cosmetic and therapeutic interventions.

It's a convincing argument in many ways, but also a tricky one, since it depends so entirely on the individual's relationship to his body and face and on the social context that frames that relationship. That's a highly idiosyncratic basis for medical intervention, and while biology also differs from person to person, it is perhaps more stable and more predictable than the psyche. At the heart of this approach is a fundamental objection to the dualism between body and mind that would separate treatments as relating to one or the other. Critical disability studies, theorized by Edmund Coleman-Fountain and Janice McLaughlin in a series of articles, supports an understanding of bodily difference and its functional and emotional stakes as being highly individual and rooted in personal experience.[37] Thomas Abrams's recent work in disability studies grounds the question of difference in subjective lived experience as a way to understand the cultural production of disability. His phenomenological approach insists on the embodied experience of the individual as a way of understanding how people make meaning of their disability. This move destabilizes the sharp distinction that early disability studies drew between biological impairment and the social condition of disability, which, he argues, asks for a division in lived experience itself. While disability studies remains divided on the status of facial disfigurement as a disability, and equally split on attempts to suture the condition, we can turn to the approaches as a way to think about the needs of the faceless on the level of the individual.[38] But this is hard; medicine doesn't work that way.

Many of the contributors to the 2004 *American Journal of Bioethics* special issue devoted to debating the facial allograft note the irony that those who need or want the surgery least are perhaps the preferred candidates, being as they have the best coping mechanisms to deal with radical facial transformation.[39] Based on this schema, surgical modification of the body and face is medicine and indeed necessary if and only if the particular psychosocial makeup of the person in question requires it. If not needed (whatever that might mean), the immunosuppressant regime required to avoid rejection would indeed make the recipient, who was not formerly unwell, into someone less well. And this really matters. As a point of departure from cosmetic surgery, this really matters.

Critics of this approach offer alternative psychological and chemical therapies or accommodations in lieu of surgery.[40] They also note that perhaps we should focus at least some of our energy on changing the broader social context to accommodate difference rather than changing those who feel themselves to be different. Along the same lines, Heather Laine Talley flips the script, noting that reconstructive surgery and its necessity, much like aesthetic surgery, is itself socially constructed; the need lies not in the individual but the society that demands it.[41] I suggest a more intersubjective analysis, noting that facelessness or facial impairment becomes a problem, or even a disability that significantly impairs daily function, in conjunction with the social conditions that react to it.[42] And both the impairment of daily function and the social conditions themselves are highly variable. Facelessness is, in the classic Goffman sense of the word, a stigma, a visible external deformity that leads to acute disapproval, isolation, and even social excommunication.[43] Stigma is real. It's also highly contextual, much like normality itself.

But—and this must be noted—another strand of disability-studies scholarship, while sympathetic to attempts to accommodate disabilities, strongly objects to medical intervention that normalizes physical and bodily difference. Exemplified by the work of Rosemarie Garland-Thomson, this approach understands both cosmetic and reconstructive surgery as aesthetic interventions seeking to produce "the normate," a physical manifestation of the standards of normative appearance. Attempts to eliminate disfigurement and disability, in this logic, use mutilation as a way to correct deviations from the norm, rendering such deviations problems to be solved by medicine. The unmodified body is that which is abnormal, to be corrected and normalized through medical modification. Those with visible bodily difference must overcome their abnormalities as much as possible, presenting a version of themselves that is in accordance with social expectations. They have a responsibility to do what they can to overcome and minimize and maybe even eradicate the source of their stigma. They must use all available resources to become normal.[44]

Garland-Thomson's approach insists on the necessarily changing nature of the body, an inventible temporal trajectory. The fluidity of the body undermines its centrality as the seat of identity, something that people resist surgically and through other kinds of bodily manipulation and enhancement. Garland-Thomson's work unites feminist and disability studies under the model of oppression of the nonnormative, nonpatriarchal body.[45] While Garland-Thomson and Gilman stand at opposite ends in their attitudes toward plastic surgery, they share a commitment to situating reconstructive and cosmetic treatments as versions of the same underlying principles. Garland-Thomson, for example, explicitly objects to attempts to erase time and history from the body as a move to erase individual identity and selfhood.[46] Both would object to the legitimation of one and the vilification of another, and, in that, both scholars stand against traditional divisions between the reconstructive and the cosmetic.[47]

The face transplant may seem like a more obviously virtuous kind of facial modification that therapeutically benefits every recipient. Such a narrative is vested in a difference in kind rather than degree between elective cosmetic surgery and the facial allograft. But there is no universality of desire for this treatment, as we shall see in chapter 4; there are potential candidates for the surgery who opt out because they neither need nor want such radical reconfiguration of their basic selves. In Garland-Thomson's terms, they resist the pressures to normalize; the more phenomenological approach might ascribe this demurral to a lived experience that does not include the manifestations of disability, or does not include the manifestations of disability to the point of incompatibility with daily life.[48]

The face transplant is not like plastic surgery in that it presents more significant medical risks, among other differences. But can we think with these scholars to consider the ways in which it is the same, bringing therapeutic benefit to those specific individuals whose experiences have dictated that they need it? The rhetoric of medical and therapeutic necessity is a powerful tool; as I show in subsequent chapters, much is gained in the institutional and public sphere by leveraging this power.[49] The narrative of medical need offers a powerful way to shift discussions and perceptions around the elective nature of a procedure. Consider, for example, the discourse of need in bariatric surgery and other weight-loss interventions; as obesity becomes defined as a medical epidemic, surgical intervention is the cure. Blame for the condition is more specifically applied only to those who don't seek treatment rather than to everyone.

But fatness is hugely fraught, as is the question of its status as disease, as a threat, and as something that needs to be fixed or treated at all. At stake are not just medical questions but questions of identity, self-image, and bodily difference. April Herndon's studies of fatness point out that the medical narrative of fatness, exactly like the dieting narrative, negates the identity of women for whom fatness is neither temporary nor personally problematic.[50] In the end, medicalizing obesity is just another form of stigma, and maybe even a harsher one. Medicalization makes fatness a more acute problem, further underscoring normalized body standards. Perhaps like facelessness, fatness on its own may or may not be a problem; the experience of fatness in a normate world may be a disabling one.[51]

The medical model does present affordances, standardizing though they may be. Bariatric surgery isn't, in this framing, cheating, shortcutting the requirement to work hard to lose weight. We can draw on Radhika Parameswaran's work for another approach to the framing of acceptable self-modification. She argues that it may be considered a virtue to improve one's appearance through labor and effort; at least, unlike those who accrue benefit through effortless and natural beauty, one has worked for it.[52] In her formulation, surgery is acknowledged as a kind of labor, a marker of those who work the hardest. Beauty surgery is no shortcut. As Imani Perry has claimed, beauty work can be highly strategic, accruing individual benefits even at the cost of broader collective identity concerns.[53]

Perry focuses on skin-lightening technology in particular, which poses individual health risks and undermines attempts to expand notions of beauty diversity. She pushes against a critique of those engaging in skin-lightening practices, pointing out the discrete advantages to lighter skin, while acknowledging the structural discrimination that makes this the case. A similar case can be made for extreme plastic surgery; in addition to the physical hazards of such interventions, they serve only to perpetuate the broad societal valorization of beauty and bodily essentialization.[54] Better looking people get more. Thinner people get more.[55] Rather than seeking to change the system, such practices honor it. As Linda Colley has described in detail, revolutions are avoided by enrolling potential opponents into the mainstream.[56] So too with beauty work, perhaps: extreme plastic surgery shows what people are willing to do to join the ranks of the insiders. The normates, maybe. And, returning to Perry, participants in these practices are not wrong to imagine the benefit they will accrue; while these interventions do not solve all of their problems, they certainly do help with some.

Is extreme plastic surgery another stage in the continuum that Gilman offers? Ought we to consider it as therapy alongside more modest interventions? There are nuances to different kinds of appearance modification, as Parameswaran and Sender highlight: self-fashioning and manipulation are increasingly acceptable as self-improvement and in the case of weight loss highly prized, but extreme interventions to the face cross the line. Especially if we can tell they have been done. Witness the furor over Renée Zellweger and Uma Thurman, both of whose faces looked noticeably different in public appearances.[57] The outcry was not that they had work done on their faces (which Thurman has denied) but that it was so obvious, and so altering. Had they followed the lead of countless celebrities to use cosmetic surgery to look like themselves—a younger version of themselves—that would have been entirely accepted and acceptable. The problem here was the transparency of the attempt to look different. We expect our celebrities to always look the same. And we also don't understand why they would want to change; those very faces had been a huge part of their success. The calculus of need here is also vexed.

But we can also resist the purely medicalized narrative of need: critical disability studies offers a way to conceptualize the individual bodily experience in conjunction with social and cultural institutions that shape pain, limitation, and identity. The rise in acceptance of altering one's appearance to conform to social norms has different pressures in different contexts: the Victorians, as Richard Sennett has shown, grew to expect performativity of the self in public because of the changing nature of private space.[58] The early twentieth century saw the rise of a heroic medical narrative that "fixed" congenital and other deformities and differences, and our current neoliberal moment celebrates self-promotion and the public performance of self, as long as the labors involved in these promotions and performances are visible and on display.[59]

The nuances are different for different kinds of manipulation—makeup is, these days, hardly in the same category as a nose job, which is distinct still from extreme facial reconstruction à la the extreme plastic-surgery reality show *The Swan*. But is the difference one of degree of suspicion, or a fundamental divergence as to the nature of these changes to the face? Do the designations of good and bad, virtuous and wicked still apply, even outside the reconstructive/cosmetic designations? Are there better and worse, or acceptable and nonacceptable, forms of cosmetic surgery? Gilman and Davis would argue that it is possible and indeed necessary to read therapeutics into these cases as well. (Which isn't to say, of course, that some interventions are not pathological. But pathology does not always, or indeed usually, align with the extremity of the intervention. Sufferers of body dysmorphia are an exception; as David Sarwer and Canice Crerand have chronicled, in these cases it is the frequency and number rather than the kinds of procedures that mark a problem that can't be solved by surgical intervention.)[60]

I bring in the extreme example to frame the consideration of deleterious health effects, which are significantly greater in face transplants than in most conventional plastic surgery. The therapeutic argument would claim that mental-health benefits outweigh any potential physiological drawbacks; this calculus is more straightforward in cases when the risk is relatively small. There are of course risks associated with surgery, and the more extreme, the greater the risk, both physically and psychologically.[61] Both extreme plastic surgery and face transplants push the question and reframe it, causing us to consider exactly how big the mental health benefits are. And that question must be evaluated on a case-by-case basis. But, in a way, the risks associated with extreme plastic surgery, some of which are posed to mental health (as the presence of psychologists on these shows attests), suggest a continuum between kinds of facial surgery: facial manipulations can and do change one's sense of self. Sometimes for the better. Sometimes, as the risks highlight, sometimes not.[62] Again, it all depends on the individual.

The labels of reconstructive and elective or cosmetic are set up as a kind of false binary between necessary and unnecessary, good and bad, virtuous and nonvirtuous. Sandeep Kaur had reconstructive surgery; Renée Zellweger made a "shocking decision to have plastic surgery."[63] But most reconstructive surgeries are elective in a way; the difference in designation affects (among other things) insurance coverage, alongside more abstract notions of need versus desire.[64] Treatments of congenital defects and other disfigurements are almost always considered necessary and thus, reconstructive, unless, of course, we imagine them to be indexical, in which case the designation is much more fraught. We don't expect babies born with cleft palates or other facial birth defects to make the best of their congenital lot in life if we can possibly help it. The success of two major international cleft palate charities, Smile Train and Operation Smile, demonstrate people's willingness to devote resources to correcting these birth defects. We don't subject burn victims to a lifetime of exposed tissue and scarring if we can possibly intervene with grafts and other reconstructions. It isn't considered vexing the index of the face, eradicating the evidence of history, changing the visual narrative, for Sandeep Kaur and other veterans of facial reattachment to get back their very own faces. On the other hand, as I've noted, debates rage around parental insistence on cosmetic surgery to remove visible evidence of Down's Syndrome.

There has always been a limit to our tolerance, and there have always been procedures that have crossed the shifting line. There is some logic to the frameworks we use: today, a narrative about reclaiming the true self, or making the outside reflect the truth through rigorous self-improvement practices, or even transforming the self through these labors, frames those practices that fall within acceptable limits. Almost always. In some ways, radical plastic surgery is the next frontier. Face transplants, despite growing tolerance for the procedure *in the extreme cases in which it is declared medically warranted* (because the virtuous/nonvirtuous distinction still exists, despite the scholarly reformulations) are not simply a difference in degree.[65] While they bear many similarities to traditional plastic surgery, there is also a fundamental difference in kind: face-transplant surgery is not simply a matter of changing appearance. It is taking on the appearance and possibly the correlated personality of another, rather than the best version of the self. And that, with its associated questions about the location and meaning of human identity, is rather a different proposition. It's not just a question of thinking the transplant as both mental and physical therapy, though that too must be incorporated as a way to negotiate the categorical complexity of the surgery. It's finding a way to think about the link between a specific face and an individual character, considering the subjective experience of each candidate, for whom that link might differ wildly. And for whom a new face may radically reconfigure the self. Or not. This is a new way of thinking the self. But we have a framework for some of these considerations, though we may have forgotten: we can turn to the early narratives around whole-organ transplants to see a very similar set of discussions about what, exactly, is transplanted alongside bodily matter. Face transplants may well be a radical, uncategorizable kind of medicine, and so we have constructed frameworks around them in order to make sense of what they confer. But, in fact, much like the therapeutic need they may address and the therapeutic effects they may have, the question of what face transplants confer must be considered on a case-by-case basis. This isn't medicine for the masses; it's treatment for the individuals. And that's hard to make sense of.

RECONSTRUCTION AS RECLAMATION (OF SELF) (IN BRIEF)

I don't want to give the impression that cosmetic surgery, unlike facial allografts and whole-organ transplants, poses no health risks.[66] There have been bioethical questions associated with elective cosmetic surgery that encompass the potential dangers of surgery generally compounded by their so-called unnecessary nature.[67] This is, however, an anachronistic framing; cosmetic surgery far predates the birth of bioethics as a field, as we'll see in the discussion of whole-organ transplants. What I want to explore here is the social stakes for facial modification in a given historical moment and how that has contributed to the changing attitudes toward acceptable self-manipulation. I'll have to be somewhat brief, and I'm going to confine my analysis to Western instantiations, despite the rich and meaningful practices of self-modification across the globe. I'm going to focus in particular on those moments that highlight the meaning invested in the face and how that meaning is exposed by forms of resistance to its modification. But, as the syphilis (noseless) example from above highlights, it works both ways: sometimes changes to the face are themselves revelatory of behavior, character, and personal history.

Any kind of plastic surgery is an intervention into the imprint of history on the face, particularly those surgeries that seek to erase the effects of time. As designer Diane Von Furstenburg recently said in an interview with *Vogue* magazine, "My face reflects the wind, sun, rain and dust from all the trips I've taken . . . why erase it?"[68] This is a minority opinion for those in the public eye. It is worth noting that history is the most acceptable kind of facial erasure, paralleling the reattachment narrative of restoring the patient to her former self rather than replacing her with someone else's history entirely. Though, as we see in the next chapter, attempts to rewind history also rewrite it.

Efforts of rejuvenation are so ubiquitous that it is hard to imagine a moment when that was not the focus of facial manipulations, surgical and otherwise. But of course medicine and malformation (and the military) have always driven research and applications for plastic surgery.[69] More specifically, battles and wars have marked major developments in this field, up to and including facial allografts, as we discuss later in this book.

Modern pioneers in the field include Sir Harold Delf Gillies (1882–1960), whose patients were World War I veterans suffering the effects of chemical and trench warfare through major trauma to their faces. This is another example of acceptable and even laudable restoration; injured vets, as Carolyn Marvin discusses and I explore in depth later in the book, must be made whole in order for war to serve its function as a social and national unifier.[70]

We have already encountered noselessness as a sign of syphilis, and for which a prosthetic (or surgical intervention) subverts its communicative veracity and experiential meaning. There were all kinds of punitive reasons that people lost their noses outside of venereal disease; as Giorgio Sperati has documented, nasal amputation was a frequent judicial punishment for adultery and other sexual misdeeds dating back to antiquity.[71] (And, it should be noted, through today: as recently as 2013, a young Afghani woman made headlines for suffering nasal and aural mutilation for attempting to escape an abusive marriage. She has since had extensive reconstructive surgery.)[72] Sperati suggests a psychosocial explanation for attempts to disguise the results that compounded the obvious practical issues, arguing that these patients felt a deprivation of their own nature and attempted to disguise it through surgery and prosthetics.[73] While the loss of a nose (and, indeed, a face) may well constitute a challenge to one's sense of self, it is hard to disentangle the emotional motivation for prostheses from the historical practical need to function without being branded a sexual deviant or villain. Broader social concerns are deeply imbricated with one's relationship to personal identity.

Noselessness isn't the only example of an ambiguous physical signifier, but it is certainly the best, not accidentally. Noses are powerful and highly visible features that are often the first part of the face to be seen, and are in many ways the easiest to remove or manipulate.[74] They are also the easiest part of the face to which a prosthesis may be attached, often dramatically changing appearance. (Consider the role of the false nose for actresses in particular who are "playing ugly," which is to say attempting to be serious and win an Oscar. Nicole Kidman in *The Hours* and Charlize Theron in *Monster* leap quickly to mind.) There are other cases in which the mutilation of the face encodes important personal and historical information, and its reconstruction changes the story. The Hebrew Bible mentions the ancient practice of piercing the ear of a slave-by-choice; the earhole represents a particular decision and life condition, and its eradication would obscure that information.[75] This is similar to branding and other forms of tattooing, and similar to the noselessness example in that it could have multiple causes and narratives associated with it.

The multiple motivations for prostheses or reconstruction cannot be isolated. But it is the desire to avoid the social stigma and practical challenges associated with a particular mutilation that ties most directly to the narrative of cheating that we see repeated in more modern manifestations of facial manipulation. The point of a nasal amputation was to brand someone a criminal of a particular kind and force that person to be identified and treated as such. Reconstruction or a prosthetic nose subverted precisely that goal, leading to deep suspicion of those with prostheses. For example, as I explore in my work on physiognomy, those with false noses found particular challenges in finding employment in service; their nasal prostheses, if detected, led employers to wonder what else they might be hiding.[76] On the other hand, if successful, their prostheses stopped prospective employees from being dismissed or rejected due to their character-damning (if possibly innocent) noselessness.

But (as you are doubtless thinking) not everyone who suffers from mutilation was or is a convicted criminal or a sexual deviant. (And, of course, not all convictions are just, and deviance may well lie in the eye of the beholder and perhaps should not be made visible to all.) Noselessness may be congenital, the result of certain kinds of cancer, accidental, or the effect of vicious, violent, and predatory attacks.

The history embedded in the face sometimes lies.

In a reversal of the narrative that facial correction is cheating, manipulating the message for personal gain, sometimes these corrections are themselves an attempt to regain visual veracity. Noselessness and other forms of mutilation may be revelatory of someone's character and his past, but sometimes these markers tell a false story. The mutilation is the cheat, and the reconstruction is the (imperfect, faulty, but nevertheless better-than-nothing) antidote. People are reclaiming themselves, or the selves they were always meant to be. Which is another way of saying, the face itself sometimes lies.

I've offered anecdotal flavor to ground the discussion about facial cheating in concrete examples. By placing these three kinds of manipulation—makeup, prostheses, and surgery—in conversation, I highlight the multiplicity of ways one can manipulate the face, and, equally, the extent to which historical resistance to these practices has always been flexible and context-dependent. That which was considered excessive manipulation was dependent on the presumed message in the nonmanipulated face. That message, and its force and strength, was itself variant over time. Equally variant was what counted (and counts) as the true or accurate external representation of the self—the before or after manipulation. In other words, when is manipulation enhancement, and when is it reconstruction of a truth that was formerly obscured? This is an individual question, and a societal one; the damaged face may, usually does, cause harm to the person in question. It may also challenge the structures of society itself, forcing into relief the question of how society treats those who are different, how to consider the ethics of obligation, and what role bodily essentialization plays in our relationships and judgments. And, of course, what the source of the damage is and what role the collective played in bringing it about. (I think here specifically of vets injured in the line of duty or afterward, battered and acid-damaged women, and survivors of structural violence and oppression.)

The easiest thing to do is ignore the faceless. Or cover them up, or fix them as quickly as possible.

So there are competing concerns when it comes to fixing facial damage. Ambiguity, again. Resistance to categories, again. There is the cheating on the one hand. The reclamation of the self on the other. And, underlying it all, there is a social need to normalize that which is different, that which challenges the structures on which society is built.

I now chart the status of the facial outsider and the perceived meaning of appearance-based difference, which directly affects the stakes for intervening with that difference. This discussion draws from disability theory, as well as the history of teratology, the history of the freak show, theater, and the performance of self, and the recent history of medicine. To start at the end, the modern makeover narrative, which includes plastic surgery among its possible interventions, posits a mismatch between outer appearance and inner character.[77] The logic of this approach dictates that the intervention is a corrective that aims to capture and display the true self, which has been obscured. The face lied, and is corrected to tell the truth. There is continuity between this narrative and the discourse of reconstruction for those like Sandeep Kaur and other survivors of accidents and attacks to the face. They were returning (imperfectly) to who they once were. Makeovers take those undergoing them to whom they were meant to be, or always already were on the inside.

The idea that there is a problem with the body that needs to be fixed, and that the way to fix this problem is through modern medicine, is neither obvious nor naturalized. It's a position. It's the bodily narrative—the body essentialization—of the twentieth century, which situates difference as a medical problem and doctors as the heroes who can fix it. We can think this in many ways; Foucault talks about biopower and governmentality as a way into the policing of the normative exterior.[78] Haraway uses the potentialities of medicine to subvert and empower bodily expectation, which finds a kind of futurist utopianism in Dan Goodley's vision.[79] The story is grounded nicely by José van Dijck's chronicle of the growing medical and heroic focus on the doctor at the expense of the patient, increasingly and explicitly situated as potentially fixable.[80] The problem, yet again, is the mismatch between the health of the interior and the nonhealth of the nonnormative exterior, which must be adjusted accordingly. We are used to this narrative, and it is hard to think of, say, conjoined twins as anything other than those not-yet-separated, those about to be saved by the miracle of modern medicine. But separation, aside from not always having been technically possible, was not always obviously necessary.[81] It changed, and change is once again afoot.

What medical authority has claimed as obvious with respect to our bodies and their normalization has a rich and variant history. It matters for our purposes because this history is shared with narratives about difference, and disfigurement, and their value and meaning. Today someone who is faceless is a problem to be solved either by medicine through reconstruction or transplantation; by internal transformation or acceptance through therapy; or by imposing barriers (masks or seclusion) to shield others and protect the self.[82] We'll see all these responses when we meet our patients in chapters 4 and 5. And they are all valid responses, all legitimate ways of coping with a painful life-changing event. Some offer more hope, and more chance of minimizing the life change (should that be desired) than others, but they are all fair choices that I do not wish to critique or attack. What is missing from this perspective is the notion that perhaps the patient is not necessarily a problem to be fixed but a person to be accommodated. As indeed all people are, though the lived experience of some in their complex interactions and negotiations with the world invite more accommodations than others.

This is neither a normative statement nor a theory of best practice. It's a historical statement: people have reacted differently to difference over time. We are now in the late stages of the heroic medical era, transitioning to a new kind of self-performance that simultaneously celebrates and polices allowable difference. Rather than the uniformity of 1980s plastic surgery (think ski-slope nose) and the anonymity of average, we are in a moment in which individuality is supposedly celebrated, and quirks are emphasized rather than eradicated.[83] *America's Next Top Model* cycle fifteen contestant Chelsey Hersley had a space shaved between her teeth to create that "individuality," mirroring runway trends of successful models such as Lara Stone, Abby Lee Kershaw, and Georgia May Jaegger.[84] A few short cycles prior, contestant Danielle Evans had to fight to keep at least some of the gap between her front teeth. Styles change, but this is more than style. This is a narrative.

It's a limited one. The contestants and the successful models are broadly traditional in type, with just a few minor differences that theoretically speak to individuality but are highly scripted and easily manufactured. And, ultimately, widely reproduced. Difference is no longer difference when it becomes a trend.[85] Here too, medicine is intervening to chase and reproduce the trends. But there is another, more fundamental change that has been developing through the twenty-first century, fast-tracked by the explosion in user-generated content: celebrity, and the ability to be famous for being famous, has a much wider and more accessible audience than ever before. There are challenges to the access, and the formula remains somewhat elusive, but the narrative of authenticity (and its manufacture and construction) means that various kinds of difference, facial and otherwise, are being performed publicly on a larger scale.[86] Think of YouTube stars like PewDiePie and Jenna Marbles, but also think of shows like *Little People Big World* and *John and Kate Plus 8*. The people featured in these shows may be presented as medical anomalies, or even problems, but they are not displaying themselves to be solved by doctors. The problem still exists, but the solution has changed. The solution is saturation. The solution—the way to success in the world, the way to triumph over personal challenges and societal barriers, the way to self-transformation and making the external factors match the interior, the way to do the necessary (and expected neoliberal) labor to be the best and most acceptable person one can be, is celebrity itself.

So too with face-transplant patients, as we'll see. With a big caveat: their celebrity comes after the transplant. No one really wants to see too much of people without faces. Not without their masks, in whatever form they wear them.[87]

But we've seen this before, or a version of this before. The freak shows of the eighteenth and particularly nineteenth centuries offer another moment of the performance of difference for profit. Nadja Durbach has compared freak-show performers to actors, arguing that they were exploiting their differences to make their way in the world, which might otherwise be impossible.[88] While the status of freak performers remains contested ground among scholars, I take Durbach's point that the prevailing narrative of the freak show was a theatrical one; these people were not problems to be medically solved, but entertainments to be enjoyed (and exploited).

There is an even longer history: in the medieval and early modern eras, and even prior, congenital abnormalities were read as signs and portents, a way to communicate with and access the ruling powers of the world.[89] The narrative of difference was a magical one; the problem was with the witnesses to the signs, not the signs themselves.

So where are we today? We perform our difference, within carefully choreographed limits. We deliberately deliver authenticity.[90] We encourage self-improvement but criticize too much change. We change our minds about what constitutes too much change. We want to know about only some of the changes, like weight loss and makeup. Other changes are an open secret, acceptable because they are a kind of return to a previous self, like botox and facelifts. Some manipulations are obvious because they are easy to decode, like breast enhancement. We shield our eyes from those who haven't complied with our demands to be better by looking better—the fat, the ugly, the different. We shield our eyes from those who we believe refuse to satisfy our need to not be offended by looking at them.

TRANSPLANTS (ONLY AN OVERVIEW)

There is another set of stories about the barriers between one body and another, and how those barriers have shifted. Whole-organ transplants have always generated a level of discomfort and concern around what else, beyond medical stability, is donated and received. But most transplant recipients were always willing to accept these risks, given the alternatives. Which is another way of saying that the bioethical question of risk was, from the perspective of the patient, always outweighed by the benefits. I use the term bioethical advisedly, and, in this case, without anachronism. After all, as I show below, it was around organ-transplant experiments that the field of bioethics was born.

But it is important to pause and note that the acceptance of transplants themselves is not self-evident. As Margaret Lock exquisitely has shown, conceptions of whole-organ transplant surgery are culturally specific and, for some, highly problematic.[91] Some of this is historical and technological, such as the resistance shown to new and unproven (and, early on, very dangerous) procedures. Some of the resistance emerges from a careful calculation of the suffering that transplants might bring with them even today, particularly in the case of children. Still another kind of resistance to transplant surgery, as Lock outlines in the case of Japan, is rooted in a particular view of what life and death mean, and what happens when the body ceases to function. While the history of transplants is in many ways an international one, there are important local nuances in the conceptualization and understanding of the procedure.

Here, I offer a very brief overview of whole-organ transplants as they relate to bioethics, with an eye on the eventual realization of facial allografts. This is more of a technical history than a cultural one, tracing developments in the procedures as they relate to bioethical debates of risk and self-harm rather than conceptualizations of the meaning of organs themselves. (For that analysis, please turn to the careful and nuanced renderings by Lock and Susan E. Lederer.)[92] Cellular memory and identity transfer provide a backdrop, if a minor one, for this discussion, and I introduce them as a way to nuance the ambiguity of the operations rather than to discuss their cultural specificity. I return to these issues more explicitly in the analysis of hand and face transplants as a troubling of bodily integrity and identity. In particular, I'll think about how the narratives of cellular memory further underscore the individual nature of these procedures and the highly specific kind of therapeutic intervention they offer. I'll think through the manifestation of the transfer narrative in film and literature, concluding with a consideration of the particular kind of surgery the face transplant is: one that combines cosmetic surgery and transplant surgery while differing from both, whose medical imperative is part psychological and part biological, and whose necessity rests on the nature of each individual candidate. A surgery that helps reimagine the body and the nature of identity.

The development of transplant surgery has both a long and a short history, and it is my task to offer only an overview.[93] Long, in that records of research and attempts date back to antiquity, with a story of Chinese physician Bian Que (died 310 BCE) attempting to exchange hearts between two men to balance their wills. There are a number of accounts dated between the third and fourth century of the Catholic saints Damian and Cosmas transplanting a leg, which is the subject of numerous medieval shrines. These narratives are likely apocryphal; more reliable accounts emerge from Indian experiments in skin transplantation by Sushruta in the second century BCE. Skin autografts (moving skin from one body part to another on the same person) were done successfully in Italy for the first time that we know of in the sixteenth century by surgeon Gasparo Tagliacozzi. He attempted a number of failed allografts (transplanting skin from one person to another), lacking the mechanism and even the framework to treat rejection, which he identified and termed "'force and power of individuality."

The first successful modern transplant was a thyroid allograft performed by Theodore Kocher in Switzerland in 1883. His research into the function of the thyroid earned him the Nobel Prize in 1909. Thyroid operations offered a model for organ transplants going forward, including the first successful corneal transplant by Eduard Zirm in 1905. The early twentieth century saw numerous organ-transplant experiments on animals and with arteries or veins in humans. French surgeon Alexis Carrel (1873–1944) drove much of the research, identifying rejection as a major source of organ-transplant failure, and developing a mechanical heart with Charles Lindbergh that operated outside the body.[94] German surgeon Georg Schöne followed Carrel's research with various experiments into immune suppression, none of which yielded major results, causing the field to languish after World War I. The 1930s saw some successful skin grafts between identical twins, renewing research into genetics and immunology, which drove Sir Peter Medawar's work with burned soldiers during World War II.[95] The war itself unsurprisingly spurred developments in skin transplantation; Sir Harold Gillies and his assistant Sir Archibald McIndoe (1900–1960), focused in particular on skin grafts. McIndoe developed these techniques over the course of World War II, making major advances in reconstruction. War has of course driven many advances in reconstructive medicine, both in terms of need and funding. As we'll see, the military has been a major (if somewhat quiet) player in the face-transplant project.

The developments and setbacks in the field brought with them a number of new kinds of questions for which medicine did not necessarily have the appropriate framework. The 1926 case of Serbian Ilija Krajan highlights the landscape into which medicine was venturing, perhaps in advance of a map. Krajan was a convicted killer who donated a gonad in exchange for reducing his death sentence to twenty years imprisonment. Both recipient and donor survived, but charges were brought against the surgeon for lying to Krajan about the identity of the recipient.[96] The case highlights questions about the rights of the donor, as well as the problems of targeting vulnerable populations for donation.

These questions simmered but, absent major breakthroughs in the field, remained in the background. This was to change in 1954 in the wake of numerous failed attempts to transplant kidneys into subjects that ranged from cadavers to end-stage renal-disease patients at the Peter Bent Brigham Hospital in Boston.[97] Dr. Joseph Murray and Dr. John Merrill took a "major ethical leap" on December 23, removing a healthy kidney from Ronald Herrick and transplanting it to his twin brother Richard.[98] The twenty-four-year-old recipient lived eight more years, eventually dying of coronary artery disease and glomerulonephritis. While the operation clearly benefitted Richard, the loss of a kidney and the procedure required to remove it posed serious risks to Ronald, raising a new set of ethical questions. In particular, do donors even have the right to consent to their own harm to benefit another? Do doctors have the right to perform these operations, which clearly violate the precept to "first, do no harm?"[99] So whole-organ transplants were also, early on, an ambiguous form of medicine that both raised mental-health questions and highlighted the complexities of what it means to heal. The ambiguity continues: where once transplants themselves may have been a violation of the law in certain countries, today opting out of a transplant may well "do harm," or so some say. Consider the debates around blood-transfusion refusal among Jehovah's Witnesses, or cases of parents choosing to peacefully let their children die instead of subjecting them to risky surgeries and dangerous anti-rejection drugs. The law still wonders, on a case-by-case basis, what interventions parents have the right to refuse on behalf of their children.[100]

The issue of consent takes on an added inflection when it comes to siblings: perhaps the health of a loved one does indeed benefit the donor in very practical ways. It depends on one's definition of health. So said a Kentucky court, when ruling on an early bioethical case in 1969. Prior to the introduction of immunosuppressant regimes in 1978, organ transplants were largely conducted between identical twins, given the success of the case of the Herrick brothers. But the issue of self-harm did not abate, particularly in the case of minor children and otherwise impaired candidates. Consent was an ongoing moral dilemma, compounded in these cases by the subjects' supposed inability to grant it regardless of the situation. The question was thrown into sharp relief with potential donor Jerry Strunk, who was of age but in a state mental hospital with what we would now call developmental disabilities. The court ruled that the potential donor, Jerry, was allowed to give his kidney to his identical twin despite the risks it posed because the survival of his brother offered him direct and compensatory benefit.[101] The law took an expansive view of the question of overall health and benefit to oneself, in principle likening the organ donation to an amputation that caused localized harm but overall benefit. And maybe this is true of living-donor donations generally: benefit can indeed be accrued by gift giving, and some may argue that this is always the case. The gift is the framework under which living donations are understood, but that does not, or should not, divorce the donation from a system of benefit and obligation. Following the classic formulation by Marcel Mauss, gifts are also part of the economy of social exchange.[102] Arjun Appadurai, Mary Douglas, and Annette Weiner have all shown that gifts, even those altruistically given, are themselves an exchange implicated in commodification.[103] This understanding matters and must be true in order to allow the donation to not violate the "no-harm" principle. The giver must benefit, according to early bioethical formulations. (Though the law has, since these early cases, evolved to grant competent adults the right to consent to harm themselves.)[104]

But the debates weren't always to be about kidneys, and the donors were not to be limited to identical twins. On December 3, 1967, South African Dr. Christiaan Barnard transplanted the still beating heart from a car accident victim into fifty-five-year-old Louis Washkansky. Washkansky lived for eighteen days, ushering in an era of intense experimentation with, and attention to, heart-transplant surgery. *Time* decreed 1967 "The Year of the Transplant," with numerous attempts at the procedure and intense media attention to the results.[105] But there were no major successes, and just a few years later most major transplant centers halted heart-allograft attempts, arguing that the lack of success simply could not justify the cost or risks. Until the introduction of cyclosporin as an immunosuppressant in 1978, there was no real mechanism to deal with rejection beyond basic typing and matching. But while the heart allograft operations ceased through the mid-1970s due to the extreme risks posed by the procedure, the ethical and legal debates continued to rage, leading to a more sophisticated infrastructure to deal with these and other emerging bioethical questions.

Alongside the lack of available organs and resources to handle them, the first and greatest challenge to the emerging surgical field of transplantation (and the one that caused its moratorium) was the risk associated with the procedures. Even in the name of experimentation, doctors could not justify the certain death of recipients (and the suffering associated with the operation), which accompanied organ transplants prior to the development of effective antirejection medication. These debates were particularly poignant with respect to the "quality of life" organs, namely the kidney and pancreas, whose recipients could live—through complicated, painful, and time-consuming alternatives—without transplants. But even for the "lifesaving" organs, namely the heart, lungs, and liver, the calculus was hardly clear.

Patients, particularly end-stage potential recipients, did not always agree with this risk-benefit analysis. Of course they didn't: on the chance of being the exception, many were willing to take the risk.[106] Others were themselves committed to the cause of furthering research, and on this basis wanted to offer themselves up for experimentation.[107] The principle of overall benefit was the legal loophole used to deal with the problem of donor self-harm, but it left a lot of gaps, particularly around the question of experimentation and recipient risk. The overall benefit to society at the expense of the individual complicates the claim that the damage to the individual is compensated by the benefits that person accrues.[108]

The early days of heart transplants, before doctors declared a moratorium on the procedures in the early 1970s, saw more optimism and, alongside it, the specter of some very challenging decisions: with the scarcity of available organs, how were doctors and hospitals to allocate them? Or, what was the best way to play God and choose who should live and who should die? These potential problems were compounded by another scarcity: operations were expensive and resources were limited, a set of conditions already encountered with the development of the kidney-dialysis machine in Seattle in 1960. This lifesaving procedure cost between $10,000 and $20,000 per year, and, once treatment began, a clear moral obligation existed to offer the patient the procedure indefinitely.

The Seattle Center had nine beds.

In order to deal with the overwhelming demand for the service, the center appointed an admissions and policy committee comprised of seven anonymous community members, including a lawyer, a minister, a homemaker, a businessman, a labor leader, and two doctors who were not nephrologists. The committee had no set criteria for evaluation and worked case by case on decisions to admit a given patient. Shana Alexander broke the story in a *Life* article in 1962, igniting a storm of controversy, and it became clear that hospitals and doctors needed to develop a more coherent and universal set of guidelines for questions of access to limited resources, including (eventually) organs.[109] Part of the problem in the 1950s and 1960s was the lack of a decision-making apparatus. Were these questions for the courts? For the hospitals? For the families? For the state? The answer, of course, was all four, but it took time and institutional effort to allocate the various issues to particular domains, and even today there is slippage between them. This matrix again highlights the problems in categorizing transplants, a problem only heightened in reconstructive cases.

The dialysis machine put the question in stark relief, and there were a variety of proposals to solve the problem.[110] This particular issue was ultimately decided in Congress through extensive lobbying efforts on the part of the National Kidney Foundation and the National Association of Patients on Hemodialysis, leading to Medicaid Bill HR 1, which provides financial support for dialysis and transplantation for those eligible for Social Security coverage. President Richard Nixon signed the bill on October 20, 1972, leading people with related diseases to lobby for similar coverage, which to date has been unsuccessful.[111] The issues of access in organ transplantation compound a broader question around access to health care generally and internationally, which remains an ongoing ethical dilemma. But now we have some sort of infrastructure to deal with it. (Poorly, usually.)

The debate around dialysis jump-started a broader American discussion on the ethics of medical care that included academics, government, journalists, medical professionals, religious leaders, and lay voices. A central concern in the debate was the definition of death: the requirement to wait until a patient's heart stopped beating meant that organs were often no longer viable for transplant. Harvard Medical School convened a committee in 1968 to establish diagnostic criteria for a "permanently nonfunctioning brain." The Harvard Ad Hoc Committee on Brain Death, as it was called, published its report defining brain death in the *Journal of American Medicine* in 1968, spurring a great deal of activity in the field through the 1970s.[112] This activity consolidated with the formal foundation of a discipline in 1978 with the President's Commission for the Study of Ethical Problems in Medicine. The group succeeded the National Commission for the Protection of Human Subjects of Biomedical and Behavioral Research and operated from 1978–1983, following which it assumed various names and forms.[113] The term *bioethics* entered into widespread use in 1973 with an article for the inaugural *Hastings Center Studies* by Dan Callahan entitled "Bioethics as a Discipline."[114] Callahan outlined the challenges facing the emerging interdisciplinary field, which counted among its members theologians, philosophers, lawyers, physicians, and scientists. Bioethics has from its outset sat between a number of spaces including the law, the clinic, and the library. In very practical terms, bioethicists have been an important voice in determining not only the direction of medical research but also the recipients of it.

While the dialysis case that catalyzed the field raised the issues of who lives and who dies and who decides, the face-transplant decisions are of a different order. There are still issues of cost, scarcity, and allocation of resources. But the question of necessity is different: people who don't receive transplants don't immediately die without them. They are both better and worse off; with faces, their ability to navigate the world is made much easier, but antirejection medicine compromises their health and shortens their lives. The face transplant gets at the heart of what counts as medicine, or healing, at all, raising the stakes on cosmetic surgery with the issue of immunosuppressants. Plastic-surgery operations pose relatively little risk and do not require antirejection meds, much like replantation surgeries like Sandeep Kaur's. The working assumption is that the psychological risk of getting what you want is relatively low.

Relatively. There was, after all, a therapist on the staff of *The Swan*. If the therapeutic stakes for facial interventions are related to mental health, then there must be some sort of mental-health effect. And it might even be quite powerful, for better and for worse. Much is gained with the transplant of a face. It might not all be good, and it is certainly not all obvious. How can potential changes in identity be projected?

Here, too, face transplants are like and not like whole-organ transplants and cosmetic surgery. They are like in that in all cases, the recipient is changed in some way following the procedure; for whole-organ transplants, at the most fundamental level, there is a tangible effect on health. Most of these effects can be predicted and charted, increasingly so over time. But the behavioral effects are far more elusive. Here, too, there is overlap with facial interventions: when the body is changed, is the person changed as well? And, to heighten the question, when the body is changed through the donation of another's organ, what else may be transplanted? Can this be predicted, too? The specter of identity transfer, which was not only an unwelcome potential side effect of the surgery but also something that crossed boundaries in a way that reframed the risk-benefit analysis entirely, was unique to transplants rather than manipulations more generally. So many of the bioethical provisions for organ donation hinged on the question of overall benefit to the recipient. Identity transfer threw into relief the very question of who the recipient might be, or how she might change the very core of her identity following a transplant.

CELLULAR MEMORY AND IDENTITY TRANSFER

There is no scientific evidence to account for identity transfer or cellular memory, though the rhetoric resonates: it feels like it should be true.[115] A very few vocal proponents support the presence of cellular memory, and there is some anecdotal evidence to support identity transfer in transplant cases to a very limited extent. The first organ transplant patients did not live long enough to even examine the question (though it was being asked), but later recipients sometimes reported significant changes that they ascribed to their new organs' previous bodies. One of the initial (but not very strong) objections to early organ-transplant experimentation was precisely around this issue: people were very concerned about the origins of their organs in terms of the kinds of people from which they came. Like faces, the concern was particularly acute around hearts, which carry a strong emotional resonance and particular narratives about being the seat of feelings. Such concerns linger; in the debate around death-row patients serving as organ donors, some raised objections around the power dynamics and the impossibility of informed choice, but others were more nervous about the implications of donating the heart of a murderer.[116]

To date, some heart-transplant recipients continue to report changes after their transplants, though many seem highly sensationalistic, including the forty-seven-year-old woman who offered the (commercially savvy) headline "I was given a young man's heart—and started craving beer and Kentucky Fried Chicken. My daughter said I even walked like a man."[117] In her comparative ethnography of organ donation, Margaret Lock cites more examples in this vein. Building on the findings of Lesley A. Sharp,[118] Lock starts with the fairly straightforward claim that organ donation is transformative. She begins with those whose health changed but moves to more behavioral stories, including the experience of the pseudonymous Katherine White, who received two kidney transplants and a liver transplant, the most recent in 1994. In an extended interview, White says that "I never liked cheese and stuff like that, and some people think I'm joking, but all of a sudden I couldn't stop eating Kraft slices—that was after the first kidney. This time around, the first thing I did was to eat chocolate. I have a craving for chocolate, and now I eat some every day. It's driving me crazy because I'm not a chocolate fanatic. So maybe this person who gave me the liver was a chocoholic?" White's doctors have a more technical explanation, telling her "it's the drugs that do things to you."[119]

White continues, "I think constantly of that other person, the donor."[120] In part, White wonders if her donor indeed was a chocoholic as a way to explain her own new inclinations. But this is a piece of a larger puzzle, in which White, like many transplant patients (though not all, as Lock shows), wonders what, exactly, she received along with her organ. In fact, transplant patients—particularly face and hand recipients—are counseled to avoid thinking too much about their new organs as foreign or from another source as a way to encourage them to integrate and adjust to their new body parts.[121] They are heavily discouraged from identifying with their donors, for the sake of their own mental health, even as many imagine themselves to be reborn.[122] But dwelling on the foreign matter introduced into their systems may well be inevitable for some, particularly in light of enduring and puzzling behavioral shifts. Transplants often force recipients to reimagine the boundaries between themselves and others, and to fundamentally reconceive of what it is that constitutes who they are.[123]

Nevertheless, there remains no serious support for cellular memory or identity transfer in the medical literature, though no one would deny the very real changes that a new organ brings with it. This is true from a practical perspective in terms of health and lifestyle changes, and it is also true from a psychological perspective. The psychological implications are particularly strong for visible organ transplants, namely faces and hands. To that end, there is an ongoing medical discussion about the psychological implications of having the visible body parts of another and the best ways of incorporating these foreign parts that are seen every day. *The Lancet*, for example, published an essay in 2006 considering the challenges of visible-organ transplant and the ways that these organs could both cause crisis and lead to changes in identity. The article does not argue that the identity changes are linked to the memories or character of the donor but that the condition of having visible donations could itself lead to identity fracture and change, for the worse and possibly for the better. Which, on some level, is entirely consonant with the bioethical and legal justifications for transplant surgery, which hold that one of the possible (and required) benefits to the surgery would be an improvement in overall health, including mental health à la Gilman and others. Which might well entail a change in identity.

The *Lancet* piece, entitled "Transplantation and Identity: A Dangerous Split?" didn't see the benefits of this kind of identity change.[124] Written by an immunological researcher and a philosopher of science, the article argues against facial transplantation solely on the basis of the psychological trauma of visible foreign body parts. The authors turn first to the precedent of hand transplants, noting that "hands are an important means of action and communication," a claim that sets up their argument that "visible organs," as "components of an individual's identity" present more significant psychological challenges than other transplanted body parts. The face is the greatest manifestation of these considerations; as "[t]he face expresses a person's identity even more directly" than hands, recipients struggle with "the constant presence of another person, and even a modified expression of the recipient's personality." The implications of such a shift, the authors claim, are acute: "With the transplant of a visible organ, a deep identity shift occurs, because one's self-image is modified substantially." In the strongest possible terms, the authors posit a universal reaction to these kinds of transplants that can have significant and possibly dangerous effects: "Every graft of a visible organ leads to an identity split, the consequences of which can be very serious if the recipient does not succeed in psychologically accepting the organ and rebuilding its social expression in every day life." To this end, they add a new metric in the consideration of surgical outcomes, arguing that "a transplant can be successful if it assures not only the function of the organ, but also the rebuilding of the recipient's identity." While quite skeptical of the potential that facial transplants can indeed lead to this rebuilding, the generally suspicious article ends on a qualified positive note in a general statement that "[t]he graft of a visible organ can lead to a full expression of one's identity, making the individual aware that to be oneself is to change constantly, and to accept oneself as changing."[125] This is a broad prescription that applies not only to organ-transplant recipients, though the article does present them as a special case for identity concerns. The article presents facial allografts as being potentially catastrophic to their recipients but reserves the possibility that they could indeed be therapeutic. It leaves no doubt, however, that these surgeries entail a fundamental change in identity.

On the one hand, this makes great sense—the identity of a person without a face, the identity of someone who lives in isolation or under constant fear or in great pain or all of the above may be very different than a survivor of a face loss that comes out the other side. But, then again, it may not be. Such things are not only subjective but also dependent on individual cultural, social, and institutional factors. Everyone's relationship to the world is different. Everyone's world is—at least a little—different. In the case of the facial allograft, change is precipitated by the fact of the clinical, therapeutic, and cosmetic intervention of a new face; that the face itself belonged to another is not the (sole) source of identity shift. Such a narrative, though framed in psychological terms, yet again reveals fears about what it means to have body parts that belonged to another and, importantly, what it means to see those parts (and have them be seen) every day. The answer to that question lies very much in the eyes of the beholder (who is in this case the beheld) and cannot be universalized, much like identity itself. Again, the biology of the face transplant is fairly stable; the psychology, worryingly, is not.

It is worth pausing to note the emphasis on the *visibility* of hands and faces in assessing psychological implications. This issue ramifies in other aspects of donation, identity, and bodily integrity. The early days of organ donation saw some initial concern with cross-racial and cross-sex donations. Indeed, as an editorial in *Ebony* noted following the second heart transplant in South Africa in 1968, from a so-called "Coloured"[126] man (Clive Haupt) to a white one (Philip Blaiberg), the heart would be allowed access to spaces it was formerly denied:

> If Dr. Blaiberg completely recovers and again walks the streets of Cape Town, a most ironic situation will ensue. Clive Haupt's heart will ride in the uncrowded train coaches marked "For Whites Only" instead of in the crowded ones reserved for blacks. It will pump extra hard to circulate the blood needed for a game of tennis where the only blacks are those who might pull heavy rollers to smooth the courts. It will enter fine restaurants, attend theaters and concerts and live in a decent home instead of in the tough slums where Haupt grew up. Haupt's heart will go literally to hundreds of places where Haupt, himself, could not go because his skin was a little darker than that of Blaiberg.[127]

The piece continues to note that, ultimately, the transplant will do little to change the oppressive situation, as "It is doubtful . . . that the transplant of a Colored heart into a white man will have any positive effect upon the rigidly segregated life of South Africa."[128] While this is a political statement, it is also a reflection on how uncomplicated such a transplant ultimately was. Cross-racial and, soon after, cross-sex donations may have caused some momentary discomfort but were generally accepted by donors and the public quite quickly. The psychological counseling that urged patients to integrate their body parts and consider them as part of themselves worked to mitigate any discomfort with these kinds of transplants. Also, and equally if not more importantly, the gender and race origins of whole organs were doubly invisible: race and sex are not imprinted on the organs themselves, which were hidden under the skin and within the body.

That which can be seen is not so easily integrated. The reminders are visible in their foreignness every day.

Hands and faces of others are a rich topic for the imagination.

THE HANDS THAT STRANGLE

It seems crazy: the stuff of horror films and science fiction, the scary stories we tell at night of still-beating hearts that come to haunt, of lingering murderous limbs that will return to choke and maim. Surely the idea of cellular memory—the notion that body parts contain within them the memories and even identities of the person to whom they belong(ed)—is only a fantasy. Surely it, and narratives around transplants more broadly, are isolated to the realm of entertainment. The stakes are considerably raised—and the narratives considerably enriched—by stories of visible transplants; the foreign body, that which belongs to another, is seen by everyone every day. At the crux of these fictional explorations is an investigation into the question of identity and behavior: does character change when appearance does? Can one, does one change into the character of someone else with the donation of her hands or face? And another question, even if so, so what?

There is a small subfield of research that explores the relationship between media representations of transplants and availability of organs.[129] Predictably, the more positive the representation, the greater the likelihood that people will agree to be organ donors. More interesting are the analyses of films that explore narratives of monstrosity and hybridity, and the threat therein, through transplantation. But monstrosity is by definition always, according to Mark Dorrian, a threat, or at the very least a transgression, crossing boundaries through "the coming together of what should be kept apart; the sense . . . that something is illegitimately *in* something else."[130] So transplantation is also always, by definition, hybridity, monstrosity, and also always transgression. And also, maybe, miscegenation, all the more obvious and visible with body parts that are daily seen. The narratives of these films underscore the extent to which identity is connected to bodily integrity and how, with the disruption of that integrity, identity itself can be transferred.

The movie narrative of evil doctors and their dangerous switching experiments is so familiar that it provided the major frame of reference for the face-transplant operation, as I explore in the next chapter. There exists alongside this tradition another set of stories about the removal of body parts from their original bodies and the risks they pose. That's vague: specifically, there are many films and stories about murderous transplant organs and limbs, usually hands, that strike against their new owners.[131] They do so for a variety of reasons; some come from murderers and other criminals, and dangerous they remain, attacking or sometimes changing the characters of their recipients. Others strike against their new bodies in resentment or anger at these supposed interlopers, often against a backdrop of the corrupt actions taken to procure these body parts, thereby justifying or at least explaining that anger. These plots depend on a certain suspension of disbelief, being as they are usually science fiction or horror, but they also depend on the possibility of identity transfer and cellular memory.

In his taxonomy of transplant films and their relationship to organ donation, Robert O'Neill refers to Patrick Gonder's category of the "body rebellion films," which were particularly popular in the 1950s and 1960s.[132] Along similar lines, Kelly Hurley uses the term "body horror" to describe films in which the body is presented as a spectacle "defamiliarized, rendered other."[133] In these films, the body is composed of a "fragmentary collection of potentially rebellious units" that express their independence by attacking and even killing others against the wishes of their owners. Gonder is interested in the body as a contested (genetic) space in which subversive racial, gender, and ideological elements can evade detection and eventually express themselves. His analysis focuses on the 1950s films, which reflect developments in genetics and heredity, but he gestures toward films of the 1930s and 40s that are concerned with brain and hand transplantation. The horror and indeed monstrosity of these earlier films emerges from the combination of parts, the piecemeal grafts that create a terrible whole. Gonder contrasts the Frankensteinian model with later films in which the horror emanates from the dissolution of the body rather than its perverted construction.[134]

Ian Olney shares Gonder's orientation toward body horror as a potentially subversive genre. He focuses in particular on the ways that narratives of rebellious limbs in cinematic adaptations of the Maurice Renard novel *Les Mains d'Orlac* resist ableist systems of normalizing bodily categories.[135] Olney traces the violations of social and corporeal norms presented by the protagonist in these films, a pianist who loses his hands in an accident and receives replacement limbs from a convicted murder. These transplanted hands ostensibly go rogue and, without the protagonist's knowledge and against his wishes, murder again. The resolution reveals another villain posing as the protagonist to murder without detection, thereby confounding what Lennard Davis has claimed is a traditional alignment of disabled bodies with evil behavior.[136] That same villain, it emerges, is responsible for the murders for which the original donor was condemned, releasing the patient from his plans to destroy his rogue hands rather than live with the appendages of a killer. Even as he and his hands are cleared of wrongdoing, the protagonist is concerned enough about the character embedded in his hands to destroy them until they, too, are exonerated.

Les Mains d'Orlac was first published in French in 1920, and provided the plot for at least four films, including Robert Wiene's *Orlacs Hände* (1924), Karl Freund's *Mad Love* (1935), Edmond T. Gréville's *The Hands of Orlac* (1961), and Newton Arnold's *Hands of a Stranger* (1962). As I explore in the next chapter, the convergence of two of these films in the early 1960s is hardly coincidental; concerns over medical development and journeys of self-transformation abounded during this era of seemingly unchecked American progress.

Here, I focus on the 1935 version, which interestingly diverges from the novel in embracing the villainous nature of the original donor. In the book, the actual villain and murderer is revealed to be an orderly in the hospital, casting the medical system in doubt even as it ultimately celebrates the triumph of this radical procedure. While the transplanted hands are, in this instance, innocent, the novel reinforces the possibility that they could well have been guilty had they indeed come from a killer. Identity transfer is not challenged; circumstance and the effectiveness of the justice system are. The film *Mad Love* goes even further in supporting notions of identity transfer and reinforcing the evil-doctor narrative. In Freund's version, pianist Stephen Orlac suffers mutilated hands in a train crash. His wife Yvonne takes him to Dr. Gogol, who is obsessively in love with her. Gogol transplants the hands of convicted knife-throwing murderer Rollo onto Orlac, who does not know the origin of his new appendages. These hands, it emerges, are unable to play the piano but can—and do—throw knives with tremendous and dangerous accuracy. Orlac narrowly, and unwittingly, misses murdering his father-in-law and ultimately saves his wife from a now mad Gogol, who attempts to strangle Yvonne and is stopped only by one of Orlac's—in this case deadly—knives.

Far from attempting to remove his rogue hands, Orlac comes to embrace and even exploit their murderous potentiality to achieve a happy ending. The monstrous body is also the messianic one, demonstrating masculine strength to rescue his female lover and sustain his traditional heteronormative marriage.[137] Medicine remains the villain—in the form of both the mad doctor and the procedure itself, which resurrected the hands of a murderer whose execution ultimately failed to rid the world of his fatal talents. While Orlac's eventual acceptance of his nonnormative body signals a kind of subversion of the typical marginalization of the disfigured character, it casts in doubt the ethics of bodily manipulation and exposes a deep suspicion of its possible dangers.

The adaptation of the novel in the early days of cinema also comes as no surprise; Stephanie Brown Clark compares the cinematic gaze to the medical one, arguing that both share an obsession with the pathological and dysfunctional body, particularly in the earliest, experimental days of film.[138] Brown Clark analyses a series of early films to show the monstrous body as the site through which concerns about the liminality of all bodies and their perpetual transformation are metaphorized. She is particularly interested in monsters as a rogue technology that subverts the control of their doctors, who themselves are often corrupt.[139] In her work, Brown Clark draws on Sharon Snyder's "ideology of the physical" (which is, in essence, physiognomy, the notion that the body and its surfaces reflect the interior) to show that the monster itself must always be a corrupt entity, given its pathological exterior. She troubles this correlation, however, arguing that monsters underlie the extent to which embodiment is itself an unstable category, encompassing as it does both mental and physical realms.[140] Here, Brown Clark resonates with Sander Gilman, highlighting the inadequacies of neat divisions between body and mind, and the connections between the two. To put it another way, but a way we've discussed already, surgery on the body is always also an intervention upon the mind and the world in which it is situated. It just isn't always obvious what kind of intervention it is.

CONCLUSION

As we think transplants and cosmetic surgery together, we think them in light of the question of the self and the ways that the self can change, and the ways that we imagine the relationship between the body and the self and how that must change. The multiple potentialities of the face transplant, as mediator and determinant of the nature of the self, drives this chapter's focus on the limits to the biological approach to identity and the variance of individual therapeutic needs, expanding the continuum between mental and physical health, reconstructive and aesthetic elective surgery, the meaning of sick and well, and the management of personal choice when it comes to the face one presents to the world.

Here's something that we don't always choose: what the bodies and faces of others that are transplanted look like, and what they might bring with them.[141] This is a difference from elective cosmetic surgery, except when it goes wrong.[142] In this chapter, I locate narratives of the self and questions of development and change within the history of whole-organ transplants, the stories of cosmetic surgery, and the rhetoric of identity transfer, arguing that the self is both a shifting target and fixed point. Therein, in many ways, lies the confusion about what the face transplant actually is: elective aesthetic intervention? Reconstructive surgery of a different kind? Therapeutic medicine? Treatment for the soul by changing the body? A way to make healthy people sick from a lifetime of antirejection medication? A method of erasing the past? Unlike other kinds of procedures, the status of facial allografts (like, in many ways, cosmetic surgeries) depends on the individual getting the procedure, and the individual's relationship to her (lack of) face. But these questions posit a false distinction between reconstructive surgery, cosmetic interventions, and, in a way, transplantation technology. The surgery opens a space to understand the psychological and experiential value or even necessity of plastic surgery that frames it as both medical and therapeutic, even as the satisfaction it brings is highly individual. This is a debate, and at the heart of it are questions about the nature of self-confidence, its relationship to the psyche, the stability of identity, and the link between appearance and the self, and the self and its relationship to the world.

It's a question of identity; not just changing one's own through manipulating the face, but taking on the identity of another.

Are changes in identity based on changes in appearance always bad? Gilman and others, including David Sarwer, would say no. Not only that, they would argue that resistance to identity change predicated on external modifications is vested in an institutional medical history that marginalized plastic surgery and, I would add, carries within it particular notions about essentializing women's bodies and appearance and the value attached to physical beauty. And, also, everyone's appearance changes over time. And everyone's identity shifts over time, perhaps in tandem with facial alterations, perhaps not. We will all one day (today, tomorrow) look different. Borrowing from disability studies, which points out that, should we live long enough, we will all one day be disabled, time dictates that we all will one day encounter faces that seem not to be our own.[143] The true monsters may be those whose faces *don't* change over the course of time. Oscar Wilde certainly made this claim in *The Picture of Dorian Grey*.[144] As we'll see in the next chapter, the (fictional?) monsters are sometimes the ones whose faces do not reflect their insides. Which is, after all, the most frightening part of facial surgeries.

Debates around face transplants operated under an ostensible risk-benefit analysis, as we'll see. This is firmly rooted in the bioethical tradition and has strong precedent in the history of transplants generally. But such a calculus depends, in part, on the categorization of what, exactly, the benefit is. The ambiguity of the surgery itself, its resistance to adhere to clear medical parameters, is partly due to the ways that it is similar to cosmetic surgery. And it is partly due to the legacy of transplants themselves, which presented a murky picture of what, exactly, is gained. But it is also due to fear. So I now inject fear, and the media through which fear is negotiated, into the conversation. Fear of challenges to bodily integrity, fear of what happens when our limbs—transplanted or not—go rogue. Specifically, the fear of monstrosity and hybridity is the fear of unreliability; the fiction of bodily transparency is laid bare not by the unpredictable behavioral effects of transplanted parts, but, of often, by the lack thereof. We fear what happens when we vex the index of the face, and, equally, we fear what does not. Perhaps the face is not the transparent indicator we have imagined it to be. Perhaps it never was.

Despite the theorizations of Gilman, Haraway, Siebers, and others about the power of self-realization and the potentiality of the hybrid, fear—cinematically expressed and chronicled by Gonder and Brown Clark—resonates. The transplanted body has remained a cinematic site of horror and disgust, a canvas upon which fears of miscegenation, ungoverned copulation, and technological hybridity has been painted. And not only a cinematic site: as we shall see, the actual face transplant repeated these debates about hybridity and monstrosity. Underlying these debates is a real fear of the limits of identity and its manipulation, and the role of the face in providing the conditions of possibility therein. Or, who is the faceless other? Who is the other with the face of another? That's the next category of cinematic transplantation exploration emerging in the 1960s: the silently lurking monster whose pathological hybrid face remains invisible. The undetected monster, whose transplant reflects and at times initiates moral and mental breakdown, is all the more scary for not being seen.

Let's see what the movies have to say about faces.

CHAPTER 3

Losing Face on Film

A man without a face is free only when darkness rules the world.
The Face of Another

The future, Madam, is something we should have started on a long time ago.
Eyes without a Face

You are alone in the world, absolved of all responsibility except to your own interest. Isn't that marvelous?
Seconds

INTRODUCTION

The 2005 face transplant wasn't the first time we witnessed versions of this surgery. Scores of novels, plays, and films have been focused on the switching of faces and its consequences. It makes sense: facial manipulation relates to multiple concerns and fears about identity, doctors, progress, technology, pain, trauma, and the prospect of life not worth living. Our concerns around face transplants have always been there, ready to be made manifest. They permeate, these concerns which have been iteratively rehearsed, so that they felt familiar already when they were raised after Isabelle Dinoire's surgery, as we'll see in the next chapter. But some moments were riper than others for these cultural imaginaries. Some moments in our history brought these particular sets of issues to the forefront.

Something was going on with faces in the 1960s. More specifically, something was going on with face transplants—an as-of-yet unrealized procedure—in the 1960s, a topic filled with concern and fear and more than a bit of anger. Something was going on that had to be discussed and experienced and worked out, playfully with just that edge of seriousness (bordering on revulsion, bordering on dystopia, firmly situated in the horrific) in the medium best suited for a topic so improbable and yet so tantalizing, so fictional and yet so very visual. Something was going on in film in the 1960s with journeys of self-discovery (that went nowhere), with narratives of self-transformation (that failed), with radical representation of identity (that destabilized identity itself.) Something happens in the movies when people try to change themselves by changing their faces with those of others. And what happens is that they get punished, severely. What happens is that someone always dies.

The victim isn't always the transplant recipient. Sometimes it's the doctor. Sometimes it's the nurse. Sometimes it's everyone. It all depends on who is most directly violating the rules of the indexical face, which dictate the unique correlation between one—and only one—person and her appearance. Everyone vexing these rules is punished and compromised; some are condemned to live a painful half-life, hiding their faces from the world. But it is the ones who go farthest in challenging the 1:1 ratio who must die, for they are the ones who truly cannot live in this world, a world whose code is based on stable visual identity. Faces may change, but they always only belong to one specific person. There are stakes to this unique correlation, as this book argues. It is the premise on which our ethical system is based. When the stability of this ratio shifts, we either need a new ethical framework or we need to rebalance the system. The films I explore here—*Eyes without a Face* (Georges Franju, 1960, France), *Seconds* (John Frankenheimer, 1966, USA), and *The Face of Another* (Hiroshi Teshigahara, 1966, Japan)—choose (in violent and aggressive ways) to do the latter.

Changes to the face, as we've seen, resist easy categorization medically, psychosocially, and culturally. Altering appearance raises questions around exactly what changes are enacted, and what their effects might be. (And doesn't this very question ignore the changes we all undergo always? But some changes are more dramatic—and more cinematic—than others.) In the movies I explore here, both nothing and everything is different when the face is, and then is no more, and then is a face (a new face) anew. The journeys in these films all end up going nowhere, but there is great violence along the way, and that violence ramifies. The world is a worse place, and so are the people in it, as a direct result of trying to replace face. As a direct result of trying to ignore the passage of time and its effects, instead attempting to rewrite history rather than incorporate its lessons. Those seeking to alter their faces do not mourn, or heal, their conditions. They attempt to eradicate them. It doesn't work. Not for people, not for countries, not for history. The lesson of these films is that replacement and revision don't work.

There are a lot of similarities in the cinematic narratives I'm going to explore, even as they are all situated in different countries and different languages. They are shot within six years of each other at the dawn of a new postwar era. The immediate aftermath of the Second World War had passed, but the memories and the trauma and the effects were still being negotiated nationally and globally, even as boundaries themselves were changing. These films directly engage with questions of individual and collective obligation and punishment at a moment when film itself was being reorganized in the new media landscape.[1] Even as they feature geographically distinct settings—which matters, as I'll show—they are all attentive to the globalizing and colonial and postcolonial restructuring of the world and its communication postwar. They all think about problems of race and identity, transformation and trauma, and healing and failure to heal. Specifically, they all feature (at least) middle-class recipients and the (so it emerges) evil doctors who seem to want to save them. They all feature themes of doubling, existential crisis, and the search for identity. They all feature questions of capitalism and the marketability and reproducibility of face-transplant procedures. They all feature highly medicalized operation scenes. They all feature the breakdown and perversion of family. And, of course, they all feature death.

While face-transplant films are hardly a genre, and while scholars don't quite know what genre they are, the three films I explore here have eventually found their way into the Criterion Collection as part of the sci-fi/horror canon.[2] Which they sort of are and sort of are not, lacking certain specific motifs of these genres while at the same time establishing new ones.[3] Vivian Sobchack distinguishes between the two genres by claiming that science fiction deals with a dispassionate analysis of social chaos and the troubling of social order, whereas horror is concerned with evoking moral chaos and disturbing the natural order through fear.[4] She acknowledges that there is often slippage between the two categories, as reflected through all three of these films, which are both scary and chilling, analytic representations of the ways that vexing the social order leads to moral disintegration. They all feature a fracturing of the family (the social) with chilling consequences for interpersonal relationships (the moral). They all critique the hegemony of capitalistic and medical institutions (the social) and the terrible actions they conceive (the moral). And they all feature murder, which is, of course, a breakdown of both. Christine Cornea, drawing on Shilling and Turner, has claimed that films that deconstruct the body do so as a metaphor for social and institutional arrangements.[5] The face, however, is a very special part of the body, representing both the moral and the social order. The disintegration and manipulation of the (integrity of the) face is a challenge to the integrity of both.

The previous chapter introduced the notion of "body horror" films, which Kelly Hurley has described as a combination of fiction, horror, and suspense focusing on the defamiliarization of the body as spectacle.[6] All three movies here play with the category, challenging what Paul K. Longmore has characterized as the traditional treatment of the disfigured, disabled, disgusting, and disgusted body. Longmore argues that the disabled body always belongs to the villain, and resolution comes only with this body's disappearance through the villain's death.[7] In these films, the monsters are those who look like everyone else. And also not: the faceless, *through their attempts to change their condition*, become monsters as well. The doctors then are also parents, bringing their monstrous likenesses into being even as likeness itself is rendered unstable. We recognize these doctors and their project; scientific and medical experiments that alter basic human form into the monstrous other are a mainstay of the body-horror genre.

Eyes without a Face, *Seconds*, and *The Face of Another* have not yet been explored in depth in conversation with one another, though at least one film festival has shown them in sequence, and numerous blurbs do consider their relationship.[8] All three films focus not just on faces and face modification but also on using the faces of others in service of transformation or reconstruction.[9] And this matters: each film is clear that the transformation project is a zero-sum game. In order for one person to gain a face, another must lose one, usually violently. There are winners and there are losers. And (as the postwar world was all too keenly aware) everyone suffers. While none of them go into great depth about the technicalities of the procedure (as yet only theoretical), they all share a strong medical character and an equally strong critique of the possibilities and impossibilities of medicalized self-transformation.

In this chapter, I take us back to this moment in the early-to-mid 1960s, this moment of highly focused cinematic interest in a seemingly fictional and even science-fictional potentiality, a moment when film itself was being reimagined.[10] I stick closely to the three films, reading their themes as a way to explore the stakes for the face specifically and for its manipulation, transformation, and transplantation, as well as broader implications for the alteration of the self through the face. These films were made around fifty years before the face transplant became a medical reality with the partial transplant of Isabelle Dinoire in 2005, and yet the themes and concerns remain eerily similar. Although the procedure shifted from science fiction to science, it still, for many, remained firmly in the category of horror. This chapter seeks to explain why.

WHERE, WHEN

France in the 1960s was still a place in mourning for its identity following the Vichy occupation and its own decolonization of Algeria. At the same time that the country was hurtling at high speed through efforts of capitalist improvement and modernization, the population was losing faith (again!) in traditional institutions like technology, medicine, and religion.[11] Film critic Adam Lowenstein, writing about Franju's *Eyes without a Face*, situates the technology of the car and, specifically, the car accident that destroys the face of the (serially unsuccessful) transplant patient Christiane, as the legacy that both haunts and drives the film forward.[12] Stefanos Geroulanos deepens this analysis by situating the car as a technology that often oversteps and endangers those it is designed to help, a situation reflective of progress as a whole.[13] Indeed all three films share a suspicion of modern developments both within and outside the medical realm, reflecting (well-rehearsed) concerns about capitalism and postwar reconstruction and improvement. In *Eyes without a Face*, cars in particular and transportation technology generally become the focus of these doubts; in the victorious and prosperous American context of *Seconds* and in *The Face of Another*, set in a postwar Japan still dealing with the legacy of defeat, the suspicion rests in more nonspecific doubts about capitalist institutions and their anonymizing capabilities. In all three films, ethics are tied to the uniqueness of the individual face being connected with an individual body. All three films wonder (darkly) about the moral affordances of hiding in plain sight. On one level, this hiding occurs through masking, either with the face of another or no face at all; beyond this surface interpretation, the films are concerned about the power of anonymous institutions to hide individuals within the collective, giving the morally unstable the opportunity to become highly, and potentially undetectably, immoral indeed.

Except of course these anonymization efforts are punished very thoroughly at the level of the individual. The institutions, faceless and all-powerful, survive the demise of some of their representatives. The status quo is maintained in all films. Challenges are posed, questions are asked, evil is unveiled and punished. But, in the end, the journeys go nowhere. The radical ethics proposed by a reconfiguring of the 1:1 correlation is rejected, for better and for worse. In these cases, it is a little bit of both: corrupt and immoral individuals are destroyed, but the institutions of which they are a part endure.

I agree with Lowenstein about the centrality of the transportation motif, but I think there is another important institutional technology that brings together all three films, one that is both modern and deeply antiquated, one whose potential for corruption is matched by its ability to offer succor and balm, one whose capabilities are, it emerges, limited only by the imagination of its designers. I refer of course to the institution of medicine and its practitioners, and to the institutional motif of the operation. The question that bioethicists raised repeatedly around the facial allograft some fifty years later was no less pertinent when the surgery was limited to cinematic and other fictional representations: just because we can, does that mean we should? In this way, the face transplant acts as a metonym for modernization itself, getting at the core of all three films' obsessive concern with capitalism as a driving force for development and as the moral justification for change.

All three films operate at the level of the individual, the very specific individuals who wish to incorporate the faces of other people. They simultaneously operate at the level of the highly general, pointing our attention to the implications of anonymization and the inability to identify individuals as they become faces—mutable and unreliable—in the crowd. And we wonder about that crowd, and the individuals in it, and how to balance the needs of those individuals against the needs of the crowd. Because, again, in the logic of these movies, the needs and sometimes the life of one (at least one, often more than one) are sacrificed to provide a face for another. But the faceless are also among the dead: ought they, even at the expense of others, be allowed to rejoin the living? (Of course not. Not at the expense of others. Which is another way of saying, there has to be another way. Another way to become nondead without taking someone else's face. Another way for the individual to heal, and another way for society to make space for the healing.)

I focus in particular on the interaction between medicine, capitalism, and identification in these films, and the ways in which the impossibility of the third necessitates the punishment of the first two. But institutions cannot be punished or really altered, all three narratives seem to claim. It is the individuals who must be punished, in part as a consequence of their own attempts to challenge the institutions, in part as the inevitable effects of time that they attempt to erase, and, entirely, as a way to demonstrate the impossibility of true self-transformation, and the failure of large-scale change. Especially when that change is really an attempt to recapture the self that once was, or the self one longs to be, through the erasure of time. These are, in the end, not stories of transformation but of replacement, failed attempts to bypass healing and acceptance of self. They are stories of the need, in the wake of radical challenges to identity, to pause. Reflect. Maybe even mourn.

This is vague. I will (spoiler alert!) briefly summarize each of the films with an eye toward that which brings them together, and then discuss each piece with respect to its attitudes toward faces and identity in particular. Each of these films is worthy of its own chapter discussing its content, its importance in film history, its broader contextual framing, and of course its technical and institutional configuration.

A note about selection: there are numerous cinematic treatments of facial modification, doubling and identity, facelessness, masks, hiding, and anonymity. These include, for example: *The Invisible Man* (1933); *Dark Passage* (1947); *Stolen Face* (1952), which shares many themes with these three films, and has strong resonance with the 1989 comedy *She-Devil*; *A Face in the Crowd* (1957); *Persona* (1966); *Faceless* (1987), which pays homage to *Eyes without a Face*; *Shattered* (1991); *Sigan/Time* (2006); and *The Skin I Live In* (2011). While all of these films think about changing faces, none of them explicitly engages with transplanting the face of another actual person, though *Stolen Face* does challenge the 1:1 correlation and would be included in my analysis were it shot later. The use of another's face in facial interventions is, as I've argued throughout this book, a key distinction, and one upon which a number of the major anxieties around facial manipulation rest. Changing the face in some way always vexes the index between face and character, and may always be cheating to some extent. But substituting the face is a different story, a distinction not only in degree but in kind. It still vexes the index, and it still cheats, but it also creates a new specter of anonymity, of iterative self-representation, of communal rather than individual manipulation. It also, always, involves someone else. According to these films, using the faces of others on a large scale demands nothing short of a new ethical framework. And, of course, it requires medical intervention and not one but two participants. One of whom may never even know her face—and by implication her identity—is being used.

Face/Off, the John Woo film whose title inspired my own, does have an actual face transplant. Two, in fact, as the main characters switch faces. As I've said in the introduction, it is important to note that the transplant(s) in that film maintain the integrity of the 1:1 correlation even as they play with it, and, equally important, the procedure in the film is reversible, implying the possibility of reversing its effects on character, identification, and identity itself. Not so with my selected three films. In all cases, reversal or regression to an earlier state can only be achieved by (someone's) death.

Each film has a different kind of transplant, both medically and conceptually. *Eyes without a Face* is interested in regeneration through transplantation; Dr. Génessier uses the facial skin of others to get his daughter Christiane's own face back. She looks more or less (rather more, in the sense of better than) herself for the brief period of successful transplantation. The protagonist in *Seconds*, Arthur Hamilton, is specifically interested in looking like someone else entirely; he wishes to change his identity. *The Face of Another* is less concerned with what and who the face is; here the face is a mask to cover a scarred and unsightly visage that itself cannot be seen. Reconstruction, yes, but not of what was once there, but rather of anything at all.

EYES WITHOUT A SECOND FACE OF ANOTHER

Let me offer you some more details so we can delve into the analysis together. I'll start with *Eyes without a Face*, partly because it was chronologically the first of the three films, partly because it is the most canonical in cinema circles, and partly because it offers the closest analogue to the face transplant as we know it.[14] The film, an adaptation of Jean Redon's novel by the same name, was directed by George Franju and released in France in 1960. It was dubbed into English and debuted in the United States in 1962 to tepid reviews, though its critical reception has increased significantly over time. British film critic Isabel Quigly wrote for *The Spectator*, "*Eyes without a Face* gets my prize as the sickest film since I started film criticism."[15] Dilys Powell echoed these sentiments, writing for the *Sunday Times* that the film was "deliberately revolting."[16] The *Birmingham Evening Despatch* (April 22, 1960) asked "WHY did they let this film come in," from France, where *L'express* spectators "dropped like flies" during one particularly gory scene.[17] The film was largely ignored in the United States, though Pauline Kael was the exception both to the European critical response and the US lack of attention, writing that "it's perhaps the most austerely elegant horror film ever made."[18] In England, the one positive response to the film nearly resulted in the reviewer being fired.[19] But times—and opinions—have changed: *Eyes without a Face* is now a critically revered film occupying an important place in the development of the horror film, with serious cult and academic cred.[20]

The film opens with a shot of a woman driving alongside the river and then dumping a corpse into it. In the first few seconds, audiences are presented with two of the major motifs that run through the film: the sinister use of transportation technology and the specter of the unidentifiable dead. Except in this case, the corpse is quickly identified by Dr. Génessier as his missing daughter Christiane, who suffered a horrible facial disfigurement due to a traffic accident (when, Christiane later implies, her father was driving.) Following the funeral, viewers discover that Christiane is (sort of) alive in the Génessier mansion, a space with all the mirrors covered, echoing the Jewish mourning ritual of *shiva* in the house of the dead, and in which Christiane lives an isolated half-life. She is seen only by her father and his assistant, Louise, both of whom urge her to wear her mask (for whom, we wonder). We learn that Christiane had a fiancé, Jacques, whom she occasionally calls on the telephone, only to hang up without a word. The falsely identified corpse was in fact the body of a kidnapped young woman whose face Génessier removed in a failed attempt to graft it onto his daughter's. Louise happily performs services for her boss (lover?), to whom she is extremely grateful for successfully reconstructing her own (more mildly) disfigured face.

On to the next victim. Louise, on the prowl in Paris, discovers a young foreign student named Edna Grüber, whom she befriends and lures to the family mansion. The increasingly uncomfortable young woman tries to leave but is ultimately chloroformed and taken into Génessier's lab. In a painstakingly close and difficult-to-watch scene, Edna's face is surgically removed. It is not only audiences who witness the procedure; Christiane lies beside Edna and willingly witnesses this perverted reenactment of her own horrible experience. The face is then grafted onto Christiane as the heavily bandaged Edna, horror slowly dawning, tries to escape through a window, and falls to her death. The initial success of the surgery proves misleading as Christiane's new and ill-gotten face is rejected and removed. Christiane returns to wearing her mask and resumes haunting Jacques, this time saying his name. Recognizing her voice, Jacques reports the call to the police, adding evidence to their investigation into the disappearance of several similar-looking young women.

The police enlist the help of a young, recently arrested shoplifter, Paulette Mérodon, in exchange for dropping all charges. Paulette checks into Génessier's clinic. Upon being released by the police, Louise picks her up and takes her back to the secret lab; as she is about to lose her own face, the police arrive at the house. The silently observing Christiane finally loses faith in her father's ability to fix her face and frees Paulette. She stabs Louise at the site of Louise's own reconstruction scar on her neck, and then turns her attentions on her father. In addition to experimenting on his daughter and her unwilling doubles, Génessier also kept a basement full of dogs on whom he performed grafts and other transplants. Christiane releases the ill-treated and angry dogs, who attack, maim, and likely kill Génessier. In the final shot, Christiane walks slowly forward into the woods outside the house, wearing her mask rather than showing her face.

Like her father when he operates behind a mask, Christiane has eyes without a face, thus the title. But as one of the living dead, a ghost with a body, a funeral survivor, a voiceless person on the phone, Christiane is also a body without a soul, as may be her father. And here is one of the many doubles in the film: the eyes stand for the body, the face(less) for the soul(less). Eyes without a face is equal to a body without a soul. With no face, Christiane is no one.

Arthur Hamilton was also no one. An unhappy and anonymous suburbanite, Hamilton decides, with the help of a mysterious organization called the Company, to walk away and start over. So begins *Seconds*, an American film based on a novel by David Ely and directed by John Frankenheimer in 1966. A stinging critique of anonymous capitalism and the myth of the isolated masculine hero, the film was a box-office flop, booed at Cannes, and initially reviled by critics. Gerald Pratley has outlined the critics who called it "cruel and inhumane," while James Powers wrote in the *Hollywood Reporter* that it was a "top-drawer exploitation item." *Sight and Sound* was disappointed that it was "uncharacteristic of Frankenheimer," who had a strong reputation at the time.[21] The innovative filming techniques and superb acting saw it redeemed, turning it into a cult classic and important cinematic precedent. *Film and Filming* noted in 1984 that it was Frankenheimer's "bravest accomplishment to date," partly because of its reevaluation of the genre, and partly because of his commitment to using formerly blacklisted personnel in and on the film.[22]

Hamilton connects with the Company through his friend Charlie Evans, whom he thought had died; that friend had used the services of the Company to disappear and begin anew. Under shadowy circumstances, Hamilton arrives at a secret location for his initial interview; while he waits, he is given a cup of drugged tea and falls asleep. He awakens to a film that seems to show him raping a girl. The message is clear: keep the Company a secret and commit to its services or face the release of this film. Hamilton's initial curiosity becomes coercion: he feels he has no choice but to participate in the Company's scheme, hiring them to stage his death in a fire (complete with corpse) and give him a new life and new identity. This is ostensibly what he wanted all along, but the Company's methods give him pause; what else might the Company do to and for him after this?

Hamilton's new identity is as Tony Wilson, a much younger and edgier man who earns his living as a professional painter in Malibu. Wilson starts a relationship with Nora Marcus, a woman he meets on the beach, and together they engage in the bohemian pleasures Hamilton had imagined he always wanted. Except it turns out that maybe he didn't, really; a new identity, a new face, and a new life don't actually make him a new person. At a party, Wilson gets drunk and begins to talk about his old life to his neighbors, in violation of Company policy. They are all, it emerges, clients of the Company, reborn to a new life. Even Nora is a plant, an agent of the Company sent to ease his adjustment. Wilson's growing disillusionment drives him to investigate his own past, meeting with his bereaved wife to learn about the failure of his marriage. Posing as an old friend of Hamilton's, Wilson discovers that his drive to material success led him to ignore his wife and their relationship. These experiences cause Wilson to radically reconsider his values, a process he hopes will lead to more success in another rebirth through the Company.

There's a catch: in order to be reassigned yet again, Wilson needs to offer new potential clients to the Company. Wilson refuses, knowing as he does that rebirth doesn't mean removal of old problems and at the same time brings with it new ones. He refuses to name names, a particularly powerful gesture in the post-McCarthy moment in which the film was shot.[23] While waiting for the next stage of the (re)rebirth process (sitting, in a particularly Kafkaesque scene, in a generic holding room that evokes the endlessness of waiting itself), Wilson encounters his friend Charlie, who, it emerges, offered Wilson's name a way to earn his own reassignment. He, like Wilson, failed in his new life; like Wilson, he thinks he has learned enough to succeed this next time. Unlike Wilson, Charlie failed to learn the lesson that rebirth ought not be imposed on everyone. With Wilson, Charlie earned his next life. Wilson, however, earns nothing through his refusal to recruit another client. When his name is called, he steps forward willingly, assuming he is to be granted another identity. In a way, he is right: he is about to become the corpse for another Company client starting over. Realization dawns as a priest recites the last rites; Wilson lies helpless on the table as the surgeon leans over and notes that Wilson was his finest achievement. And then not only Wilson but also Hamilton die—this time forever.

In *The Face of Another*, a Japanese film from 1966 by Hiroshi Teshigahara that was based on the Kobo Abe novel of the same name, the search for the right kind of face, and the person with the right kind of face, plays a more explicit role. And the right kind of face is the one that would act as a good— which is to say unremarkable—mask for the protagonist, Okuyama, who suffered disfiguring facial burns in an industrial accident. Okuyama spends his days swathed in bandages that form a barrier between him and his intimates, including his increasingly distant and discomfited wife. In meetings with his psychiatrist, Hira, Okuyama agrees to an experimental (and, ideally, monetizable) procedure involving a highly realistic mask indistinguishable from the face upon which it is modeled. Unlike the other film narratives, however, in *The Face of Another* no one dies in service of procuring a new face; death comes later. The third of four collaborations between Teshigahara and Abe, the film was well received in Japan but was rather less successful internationally. Nöel Burch criticized its unfashionable modernist style as an "extravagantly chic . . . abstruse exploration of psychological symbolism."[24] While still less well known than the other two films, *The Face of Another* has had a modest critical renaissance along with the duo's higher profile *Woman in the Dunes* (1964) and other collaborations.[25]

Okuyama is able to don and remove his masks, pivoting between his (lack of) life as a bandaged outsider and his identity as the man masked with the face of another. As Hira cautions, the mask begins to take over his personality, causing him to act in uncharacteristic ways. Okuyama tests his new face secretly, renting an apartment while "away on business" and attempting to seduce his wife in his new skin. She easily acquiesces, sparking Okuyama's anger and sense of betrayal over her infidelity. She expresses surprise as she had known his true identity from the outset, playing along with what she thought was a masquerade designed to bring them closer together and bolster their flagging relationship. Disgusted by her husband's attempt to entrap her, she leaves and locks him out of their home.

Shaken and abandoned, Okuyama wanders the streets aimlessly with his mask intact and mostly in control of himself. But only mostly: he lashes out, assaulting a woman, which leads to his arrest. The police find Hira's name and call the psychiatrist, accepting his explanation that Okuyama is a harmless, if disturbed, psychiatric patient. Hira comes to collect Okuyama, who tries in vain to assert how dangerous he is as a way of denying his status as mentally disturbed. His objections only reinforce the characterization, and he is released in Hira's care. As they leave the station, they walk against a crowd of faceless people, underscoring Hira's fantasy of an anonymous and ultimately unaccountable society, facilitated by his masking technology. Also fuelled by Hira's masking technology? His own stabbing, at the hands of the now deeply fragmented and unstable Okuyama and his (Hira's, Okuyama's) mask.

The Face of Another features another death; in a parallel secondary plot, a woman with disfiguring burns on one side of her face suffers from debilitating loneliness due to her appearance. She tries to reach out, desperate for contact and human connection, to the only person who she thinks can love her in her condition, her roommate and brother. Horrified by her transgression (all we see is a kiss, and we are left to wonder), she walks into the ocean with her distraught brother witnessing her suicide from the window. Unlike Okuyama, she is not given the opportunity to reclaim life, identity, and personhood with a new face. She's never even given a name. (No woman is in *The Face of Another*.) Rather than exist in limbo, she chooses to die. Her death, though framed as an act of contrition, is not a punishment but a statement of the impossibility of her life. She gives us the opportunity for compassion, if tempered, for Okuyama, while at the same underscoring the necessity of the attack on Hira. Unlike hers, Hira's death is a punishment. He had to die. Like those who challenge the index in all of these films, he had to die.

These are three different films from three different countries in three different languages, all clustering around the same time. They have a lot in common, and a lot that divides them. Of course, they are all interested (obsessed) with the substitution of one face for another. What sets them apart from the myriad films interested in facial transformation is the very specific nature of these alterations; they all deal with the substitution of a face with one belonging to someone else. These are not films about plastic surgery and facial modification and improvement, or even about different faces for different personas (à la Jekyll and Hyde). Each of these films seeks to erase or hide a damaged person by layering on the face of another person entirely. And that very specific kind of transformation has particular stakes beyond those of facial modification, beyond the vexing of the index provided by the face, beyond the claims that someone who changes his or her face is cheating the dictums of nature and fate. (All of the transgressions just mentioned have implications, of course, and I have explored some of them and will touch on others.) Face transplants have all of these problems and more: they change one person at the expense of another. They change the landscape of knowability and the rules of interpersonal engagement. As such, they demand nothing less than the reorientation of the ethics of human interaction. These films take on the morally fraught arena of unwilling or unknowing transplantation, ultimately punishing those involved in a very exact kind of quid pro quo: if you steal the face—and thus the identity and personhood of another—you must die.

The *you* in that sentence is deliberately ambiguous. It isn't, always, the patient who is the thief, and so it isn't always the patient who dies. Without a face, these films argue, one is without a true life. And these already dead faces of people can't—or shouldn't—die again. When they do, as in *Seconds*, it is to restore balance when it has been terribly disrupted.

DOCTORS, OPERATIONS

This time, I'll start with *The Face of Another*. It's not first chronologically, it is the farthest away geographically from the other two, and it is in many ways the outlier. (Though, in fact, each film is the outlier across one vector or another.) The most obvious difference between this film and the other two is that the face donor here is only a model, selling his likeness for use in a mask rather than for transplantation purposes. As a mask, the operation here is reversible, easily removable, and not really a surgery at all. Okuyama, unlike the other patients, rotates between his identity as a man with a face and his life (such as it is) without one. And, yet, his bandages serve as a mask perhaps as much as his borrowed face, altering his interactions with others and filtering his own emotional experience. As his face mask slowly takes over his identity and character, the ability to remove it seems incidental; Okuyama becomes his masked self rather than the mask allowing him to once again live his life as himself. Though, like in *Seconds*, returning to his old life was never the goal.

While *The Face of Another* bears least resemblance to modern face transplants as such, the donor-patient relationship is perhaps the most similar. Of course (unlike some other organ donations), face transplants, which involve actual facial tissue, require a nonliving donor, and in *The Face of Another* the donor is both alive and not entirely aware of the purpose for his visage. But, in contrast to the other films, there is some kind of consent in the exchange; no one is kidnapped, manipulated, blackmailed, or violently violated to give or give up his or her face. No one is kept hidden and isolated to be an experimental subject, a kind of lab rat at the will of an overarching controller. (Though like Christiane, Okuyama is in thrall to a powerful doctor figure, or a Svengali.)[26] So perhaps this isn't the story of a face transplant as such, but it certainly qualifies as the story of the face of another (as the title makes clear) and, specifically, the power and powerlessness that comes along with trying on someone else's identity. Though Okuyama could take this identity—which is to say the face—off, something is always left behind. The face of another changes you. It changed him. He couldn't ever completely take it off.

No such removal is possible for Wilson in *Seconds*, even when that is what he most desperately wishes for. His original identity, as Hamilton, was ostensibly easily (if expensively and laboriously) replaced, but, as Wilson, he's stuck. Here lies one continuity between the two films: it is harder to shake an artificially assumed identity than the one with which one is born. Which actually means that the act of taking on another identity (via another face—in this case an artificial, constructed one) is an immutable act. It changes one, sometimes for the worse. So perhaps it is not faces themselves, or faces alone, that alter character but actions, which isn't in the end all that surprising a message. What might be more surprising is the repercussions for this particular action, that of taking on another identity. Based on these repercussions (murder, death, destruction of marriage), it is a very bad set of actions indeed. I'll talk about why that is, and why that might be surprising, below, but first I will explore the question of the donor and the donor relationship in *Seconds* and *Eyes without a Face* in more depth.

Like the other two films, *Seconds* contains a detailed and painstaking (and often painful) visual sequence outlining the process and charting the surgery, starting with classic before-and-after drawings. The expected medical apparatuses are shown, including bandages and clamps (many clamps, this is a bloody and messy affair), masked doctors, and lots of sutures. The sequence was shot in an actual operating room, and the director was so committed to *cinema vérité* that six crew members fainted filming the hyperreal scene.[27] It is some sort of radical plastic surgery that turns Hamilton into the much younger, far more dashing Wilson, combined with a vigorous exercise regimen shown through a transformation montage of physical activity. This type of surgery was within the realm of medical possibility; as of 1966, people could get all kinds of aesthetic facial interventions, and the decade's emphasis on self-improvement saw a dramatic rise in facelifts through the end of the 1970s.[28] The final scene, in which the plastic surgeon informs Wilson that his Wilson-ing was the surgeon's proudest achievement, lends support to the theory that the transformative procedure being performed is extreme plastic surgery. (There is perhaps a vague tinge of the bittersweet even in the moment of murder, as we recognize that the surgeon here is now about to murder his baby. But we quickly remember that he does so routinely; Wilson was the surgeon's best but hardly his only rebirth. And hardly his only re-death.) So here too is no real face transplant, at least as far as we know.

And yet I include these films, despite my insistence on excluding cases of facial modification, aesthetic surgery, and other nontransplant forms of facial transformation. But, actually, my insistence is on including narratives that exchange one face (or lack thereof) for another, rather than on excluding stories that don't have a medical transplant. These are stories that highlight the resistance of the procedure to categorization and raise the question of what is gained when the face is changed. Both *The Face of Another* and *Seconds* clearly occupy the space of exchange. It happens in reverse order in *Seconds*; here, the dead take on the identities and faces of the living rather than the other way around. Also, we don't know exactly how it works in terms of bodies—how does the aging suburban Hamilton acquire the body of the hip urban artist Wilson?

In a possibly infinite regress, Hamilton's body—required for a successful rebirth—is supplied by a failed reborn. As soon Wilson himself will be, giving another the opportunity for rebirth. Which likely will prove unsuccessful because, it turns out, according to the narrative in *Seconds*, only some things change with a new face. Others, the things that made Hamilton want to abandon himself (sort of; he only sort of wants to abandon himself, but, once he begins the process, the blackmail tape means he has passed the point of no return), do not die with his identity. A new face and new life may bring with it new opportunities; they don't bring an entirely new self.

Does *Seconds* differ from *The Face of Another* in this respect? *The Face of Another* argues that, with a new face, Okuyama becomes a new person, whereas *Seconds* argues forcefully that some things never, and can't, change. But the difference here is not in behavior, or behavior motivated by appearance, but rather in what kinds of changes are being sought and wrought. Okuyama had no face and, thus, no identity. He was a kind of tabula rasa, and his mask provided (some of) the writing. Wilson, however, could not leave Hamilton behind, because he never occupied that liminal space of the living dead, faceless but bodyfull, waiting for an identity to fill in the blanks. Wilson carried Hamilton inside him; his Wilson face added to but did not erase his essential Hamilton identity. Not until he provided his Hamilton body to rebirth someone else. Which he did against his will, but he is no innocent; someone else provided a body for him to be reborn. He did not steal a face or identity to become Wilson, but he did steal an identity to cease being Hamilton. That donor, a failed rebirth, was presumably equally unwilling as he to die. However, Hamilton's greatest sin, unlike Okuyama's, was not identity theft. It was trying to change himself too much.

Unlike Wilson, Christiane Génessier lives only in the liminal space. The entire film, save the brief period before she medically rejects Edna's stolen face, sees her as neither alive nor dead, yet at the same time both. Like Hamilton, she is technically a corpse; her father uses the body of another to explain Christiane's absence from her own life. However, she is not trying to escape herself. She is trying to regain herself. Or, rather, her father is trying to regain his daughter and, in so doing, release his own guilt for destroying her face and, he thinks, destroying her. This narrative implies that her morality, like the rest of her, is gone with her face; her passivity in witnessing her father's and Louise's actions in kidnapping and horrifically maiming young women against their will supports this reading. Her father's flexible morality is rather more complicated, but we'll get to that soon. First, let's think about Christiane's donors, and what it is they are (unwillingly) facially offering.

Christiane, unlike Okuyama and Hamilton, is not seeking a new face. She wants her old one back, and, in the medical world of this film, the faces of other young, similarly featured women offer this possibility through their skin. While the surgical approach in the film bears the strongest resemblance to the modern transplant in that it takes the face from another person and grafts it onto the patient, the presumed outcome is quite different: rather than looking like the donor, or some hybrid therein, Christiane looks again like herself, only, as Louise says, even better (59:08). Like an angel. (But we already knew that Christiane isn't of the living in the same way as the others.) The new face is a transplant, but in service of reconstruction rather than regeneration or renewal. Christiane seeks to rejoin the living as herself rather than as someone else. But of course she is already dead; it's impossible.

THE OPERATION

It's the most famous sequence in *Eyes without a Face*, and it is hard to sit through. At the first screening in Edinburgh seven viewers fainted during the sequence, and even audiences for the recent rereleases experienced significant discomfort.[29] In Barbara Creed's terms, this is the moment of the abject gaze, when the viewer turns away from the monstrous-feminine. The moment when the human is separated from the nonhuman.[30] The moment, in this case, when Christiane's nonhumanity is made manifest, when the nonhumanity of the women in the film is made manifest. I refer to the operation, shot in extended detail in the style of the science film, in stark silence, with careful close-ups of the surgical apparatus, the sweat on the doctor's brow, the gloves and masks and instruments, which make viewers squirm and almost long for the release of the truly horrific climax.

We watch Génessier peel the still-living and entirely unwilling Edna's skin off her face, turning her into Christiane, into worse than Christiane, more dead than living. She's asleep, sedated, but we know she will soon wake up to the horror that her sort-of life has become. It is this scene, more than any other, which has situated the film so firmly in the horror category. (Though the final scene of dogs tearing Génessier apart runs a close second.) It's not only body horror. It's sheer violence. As we watch, we know that we are not the only spectators. Christiane watches too. We watch her watching. She is unhorrified, even excited about the possibilities presented by Edna's face. Edna's faceless limbo may liberate Christiane from that same fate. It is Christiane's excitement, her lack of horror, which (in a film with no shortage of horrific moments and messages) may be the most horrific of all.

And Edna knows what is coming. She has seen it, as we do, through her eyes. In the only instance of a direct view of Christiane's unmasked face, Edna is presented with a vision of her future. But that future is not Edna's alone; as we already know from the funeral, Christiane will forever have eyes without a face. She may briefly return to the state of the living, as she does immediately following her transplant, but it is always only going to be a visit, because not only does she not have a face, she has, in her acquiescence (miserable though it may be) to this operation (and presumably others) also lost her soul.[31] She (like her father, soon to be dead) is no one.

(And here too we can turn to Giorgio Agamben to think about Christiane's father's desperation to save her, to rehabilitate her into society, to transition her from a state of bare life which, he thinks, is a damning condemnation of himself.[32] Little does he understand that it is in fact his efforts to save her, efforts that entail torturing dogs and young women, that are the true damnation. But we can also turn to Hannah Arendt, who describes the ways that being in public, and being of the public citizenry, is a political act.[33]) Christiane doesn't go in public, ever. She is not a citizen, and, as we know, she is not really a person. It is not her facelessness, or not her facelessness alone, that removes her from the populace. She hides—behind a mask, and behind walls. Getting her face back would not necessarily reinstitute her into the world. She would still have to leave her mansion-prison-childhood (home), and she would still have to allow herself to be seen. As we learn, in the final scene, she does not remove her mask.

It is not just Génessier's failure to transplant a new face that makes the damage to Christiane permanent. Her refusal, and her father's refusal, to accept her new state of being and their combined efforts to manipulate time and history changed her for the worse. In attempting to make Christiane appear less monstrous, her father participated in making her into a monster. Just like him. The moment of horror, the moment in which we look away, is the moment in which we realize the extent of the monstrosity; Christiane is not a monster because she has no face. She's a monster because she doesn't want anyone else to have one either. She will never become human, she knows that. So, in a kind of vampiric imperial exchange, she helps her father to make others like her. (But, perhaps as women, they already were always like her. The removal of the face is the unmasking of the monstrous-feminine. Is it that moment that makes the women into monsters, or lets us know that they are?)

But on to the technicalities. It is a surgical scene, charting a medical operation. The doctor looks like a doctor: masks, gloves, scrubs, nurse assistant. There are clamps, lots and lots of clamps. Génessier marks the outline of Edna's face, and then lifts off her face. It's a face-lift in two senses: a perversion of the aesthetic surgery, and a theft. Génessier is changing her face, and he is stealing it. And, as we know from the other young corpse, it is a pointless theft, destined to be a gift that quickly destroys itself and slowly destroys those around it. It is a gift that takes itself back, and takes everyone down with it.

Arjun Appadurai has troubled the distinction between commodity and gift, arguing that gifts too entail exchange in the form of obligation and reciprocity, and emerge out of self-interest.[34] For Edna Grüber, however, the face is neither gift nor commodity. It is simply not an object that she understands as one with exchange potential. She does not consent to give or sell her face, and it is certainly stolen from her, but there is another theft: taken from her is the notion that her body is her own, under her control, indivisible and unique to her. She didn't agree to give her face, or even to sell it, and she didn't agree that she could. In *Eyes without a Face*, the gifting framework is a deeply perverted one, echoing the deeply perverted framework under which the characters are operating. The owner of the face is the victim of theft, corrupting the exchange from the outset, as reflected in the toxicity of the face itself, which literally self-destructs.

Génessier demonstrates Appadurai's claims admirably: his gift to his daughter is certainly loaded with obligation and expectation of reciprocity. He wishes nothing less than absolution from her, but no gift that he can give will ever be enough to earn that. It is he who took Christiane's life, in the form of her face, away, an accidental act that he deliberately repeats with each of his victims. What restitution does he offer them? And Christiane is a perverted recipient, participating through her passivity in the theft that allows her to receive her gift.

We cannot separate the operation from the doctor figure, and from the specter of medicine writ large. All three films engage explicitly with the dangers of medicine and the risks of putting faith in doctors and their science. Science is not always progress. Science, these post-Hiroshima films reflect, can be dangerous. Science and morality do not go hand in hand, necessarily, and this film emphasizes how and why. (Even with its brief moment of humanization, in which Génessier plays with a young male patient, the film is unequivocally suspicious of medicine, science, and technology.) There are many villains in this film; there might be only villains as actual characters: Génessier, Louise, Christiane. Everyone else is just a plot device.

But why the major downer on medicine, and why the face transplant as the representation of the chilling evil medicine can perpetrate? In a way, it's obvious: the face, according to the narrative of this film (and it might not be wrong, as this book shows), is our passport to public life, public interaction, the world at large. The face is the soul is the world is life. People will go very far to keep those things intact. (Is it different now that we can have public life without faces? A question for Lévinas, and for the conclusion of this book.)

Eyes without a Face isn't alone in making these grand correlations and grand condemnations. Let's move to *Seconds* second in this section. Equally suspicious of the institution of medicine, *Seconds* absolutely affiliates surgeons with capitalist institutions; the Company and its operations are literal operations as well as financial ones engaged with rebirth and its technicalities. The Company is at once a machine and a beast; insatiable, it needs an endless stream of new clients to feed itself. And, to feed them, it needs an endless stream of old client corpses. But it has no head, and cannot be slain. The stream will dry up only if all its clients to refuse to offer new names if and when they request a new rebirth. But if they refuse, they die. The Company can only be destroyed by kamikaze, consciously or otherwise.

While the surgeons, and the medicine, are interchangeable with the institution itself, there is one surgeon figure, a limb of the beast, a cog in the machine, as it were. This surgeon performs two operations: the one that turns Hamilton into Wilson and the one that turns Wilson into someone else's corpse. We don't really see the second either, but we do see the moments just before it happens. And, in those moments, we see the familiar cues for Medicine writ large: scrubs, surgical table, face mask, instruments. And we see another thing: a patient coerced (through blackmail) and an unwilling donor. Wilson doesn't want to give up his body to someone else. He doesn't want to die. His identity and his corpse are no gift, nor are they a willing exchange. They are, like Edna's face, stolen, and, like Edna's face, intended to be given to someone else; although in this case they are bought and paid for by the next client of the company, while Edna is an innocent.

Like Christiane, Wilson is not innocent; he's abandoned his life and his family (and we'll get to family soon), allowing them to believe him dead and mourn over him. But his final moments give some truth to this lie, redeeming him from his duplicity and saving others from the same path. And while his lack of lucidity in his final, anesthetized moments limits the scope of his redemption, the final image of Wilson walking along the beach with a child—possibly the granddaughter whose birth he missed while "dead"—speaks to his awareness of what he's lost. Wilson refuses to play his part in the perpetuation of the Company system; he dies for his refusal, but in so doing he escapes from limbo—unwillingly but, perhaps, nobly. As a gesture, his refusal is ultimately useless, as the packed waiting room and the actions of his own friend (who recruited Wilson to grant the friend another rebirth) demonstrate: most others are prepared to sacrifice their friends (even knowing that rebirth often leads only to rebirth or death, not a real new life) to play their part in their iterative existence. Like Christiane and Génessier, they lost their souls when they lost their original faces. Though they seem to be alive, they too are in limbo, waiting for the death of their current identity and their next rebirth. Which isn't really a life at all.

While there is a critique of capitalism and consumption in *Eyes without a Face*, focused particularly on modernity and technological development, this narrative is far more explicit in *Seconds*. (It's stronger still in *The Face of Another*. I'm getting there, I promise.) Couched in terms of excitement, opportunity, and the benefits of the novel and untested, the product offered by the Company is rather less than the transformative experience it seems to be. What patients who are also clients (a familiar category in the United States) buy into, in the end, is the perpetuation of the Company itself by providing more and more human stock for its process. The laws of thermodynamics are at play here; matter is not made out of nothing. When people die, there have to be corpses. And for the company to make money, people have to die for real so that the clients can die for fake. Except that fake death is a very real death, the death of their integrity and their soul and maybe the best parts of themselves. They just don't know it yet. It is only at the moments of their actual deaths that they get the chance to reclaim it.

It seems like a wonderful opportunity, the chance to start over. But it is far too riddled with corruption and filthy lucre to ever work. The Company (with no other name than that, meant maybe to refer to all companies, to the company system, to the workers in the company system who keep it going) itself has no morality except the almighty dollar. Which could make it neutral, at best, but doesn't: soulless, always, the ever-hungry machine-beast also exists in limbo, forever, stealing lives and faces and souls and condemning others to its fate. The freedom it sells is an illusion, enslaving its clients to itself in turn.[35] It wasn't always this way, or it didn't have to be: in a melancholy scene toward the end of the film, the Company's founder tells Wilson that his original intention had been to offer a genuine service to those in need, but now the Company is run by "a board of directors on a profit-sharing basis." Progress doesn't always mean better. Or, as all three films teach us, just because we can, doesn't mean we should.

Doctors are an integral part of the process (progress). They make it happen. Not only that, (as we learn in the final moments of the film) they are proud of it. As in *Eyes without a Face*, the pursuit of scientific and medical success overrides and pollutes conventional morality. Ever chasing the goal of the more perfect manipulation/transformation/transplantation, medicine proves itself to be a highly faulty judge of what should be done rather than what could be done. We ought, these films say, to be very suspicious of medicine as moral arbiter. And maybe we ought to be suspicious of the individual doctors as well.

And maybe we ought to be most suspicious of doctors like Hira. On the face of it the most innocuous, the least corrupted in terms of what he does to his patient and his donor, he might be the most dangerous doctor of them all.

The Face of Another offers the most transparent (and, in the post-Hiroshima moment, most personal) critique of the connections between medicine, capitalism, and morality. Hira's motivations in fashioning a mask for Okuyama are forthrightly stated throughout the film, and they are not rooted in helping his patient find peace. Rather, he approaches the process like an experiment, albeit one that is, in his words, "against medical ethics" (22:40). ("But," he continues, "you've twisted my arm.") Rather, Hira is fascinated by the possibilities presented by eradicating appearance as a means of identification. What would a world that can't rely on faces be like? It would be different: "Masks like this could destroy all human morality. Think about it: name, position, occupation: such labels would no longer matter" (1:00:37). What would matter is how many people would be interested in having their morality destroyed—Hira suspects a great many: "I could mass produce them. A face, easily removed. A world without family, friends, or enemies. There'd be no criminals, hence crime would disappear. Unbounded freedom, hence no yearning for it. No such things as home, hence no dreams of escaping from it. Loneliness and friendship bleed into one another. Trust among people, now so richly prized, would become obsolete. Suspicion and betrayal would no longer be possible" (01:27:50).

Hira speculates on the liberation that would come with destabilizing the link between faces and specific individuals. It would destroy morality, he thinks, but for the better: morality brings with it desire for the forbidden. Without the threat of identification, nothing is forbidden, nothing is denied, and nothing is transgressive. And whoever (Hira) provides the technology to enable this way of being in the world would richly reap the rewards.

Clearly, there are contradictions to be unpacked and logical fallacies to be demonstrated in Hira's theories of morality. Let's take that as a given: it's not the point here. What is the point is that Hira—not a surgeon, actually, but a psychiatrist (as he is careful to let us know at the outset of the film)—is well aware of the destabilizing potentiality of the mask he makes for Okuyama. He understands the challenges such masks present to the collective, and he is also entirely aware of what it might do to Okuyama himself. Throughout the film, Hira frequently notes to his patient and others that the mask changes the wearer, and that these changes could lead to a deep fracturing of the self.[36] He consistently reminds Okuyama to document how he is feeling for Hira's own purposes, maybe medical but likely commercial. Hira needs to know how the mask works in order to market it most effectively. Or, at least, to make it safe(r) for the wearer (and those around him or her).

Okuyama is indeed in very real need. He is suffering. His loss of face destroyed his conception of self and his ability to exist in intimate or even casual relationships. He is desperate, and he is angry, talking with Hira about burning his wife's face so that they would look the same, so that they can relate to one another, and really so that Okuyama can reveal Hira's lie (spoken during a therapy session) that faces don't matter beyond social prejudice. Of course they do: that's why burning his wife's face is such a terrible threat for Okuyama to pose. Okuyama, as indicated by his semiserious comments about burning others and his repeated laments about being invisible and wishing for the world to be in darkness (or for him to "gouge out all of humanity's eyes," at 08:12), is already unstable before he dons the mask. But the mask makes it much, much worse. Hira, in providing the mask, makes it much, much worse.

Hira's lie about the meaninglessness of faces is almost ironic in its transparency; the entire film is devoted to showcasing the many ways in which nothing matters more than the face, for the individual sense of self, for relations between two people, and, most importantly (for the first two reasons could perhaps adjust to a change in social norms), for social structure as a whole. Changing the central status of the face would require a restructuring of all of social interaction. Hira says so himself, and he's right: the stability of the 1:1 correlation undergirds the stability of morality, responsibility, and accountability. Assuming we believe that people behave properly at least partly out of fear of getting caught. It's a shallow kind of morality, and certainly not strong enough to curb the impulses of many of the characters in these three films. But we can see why Hira wants to remove even that limitation. And we can also see how accountability to the other, to the specific other whom we face, might pose an even greater moral obligation.

There is no real operation scene as such in *The Face of Another*. Rather, there are extended sequences of a more technical nature, detailing Hira's construction of the mask and his guidance on how it ought to be applied. Here, medicine and technological industry are totally allied in both practice and motivation. Okuyama is patient, experimental subject, and test client all at once, using a commodity designed to make his life proximately better, in the short term, but which will quickly make it much worse.[37] ("I feel more like a guinea pig every day," he says at 01:27:46). Hira knows this, but doesn't really care. And Okuyama sort of knows this too, a little, maybe (and let's remember the informed consent debates here: can Okuyama really know the risks when he is so desperate?), but also doesn't care.[38] So, everyone wins and everyone loses, an inevitable outcome when medicine operates in service of money.

FRACTURED FAMILIES

Okuyama's mask, or, really, Okuyama's deception about his mask, provides the final blow to his faltering marriage, but it certainly does not bear full responsibility for its collapse. We actually don't know exactly when their troubles began; the narrative of the film starts after Okuyama's accident, and he and his wife are already experiencing significant difficulty with one another. The accident, and Okuyama's declining sense of self (worth) after the accident, offer serious challenges to their relationship, but we don't know what their relationship was like to start with. We do know that Okuyama has little respect for his wife, seeking to trick her into infidelity (with his masked self) and treating her with contempt when she (in his mind) succumbs. Okuyama may feel like nothing and no one without a face, but he does not doubt that he is the one in control of their relationship and its direction, and he doesn't hesitate to destroy it. He is wearing the mask when he seduces his (nameless) wife, but the plan had been launched well before the mask took him over; his actions cannot be blamed on the mask and its effects.

Okuyama regrets his deception, deeply, but, as his wife tells him, that was the point of no return. It wasn't Okuyama's appearance, or his moodiness, or his cruel treatment of his wife that was unforgivable. It was his attempt to trick her; it was his attempt to pretend that, in donning the face of someone else, he actually was someone else; it was his dishonesty about his face and the possibilities it presented that made her walk away for good. It was the unacknowledged use of the face of another.

Manipulating, transforming, transplanting faces destroys families. In all three films, families break down as a direct consequence of facial manipulation. And, as *The Face of Another* is careful to underscore, it isn't the effects of facelessness that raise the greatest challenges to the family unit; it is attempts to regain face.

Okuyama's relationship with his wife isn't the only family dynamic explored in the film. The secondary plot probes the fraternal bond and its limitations, and the tertiary narrative quietly reveals Hira's own familial complications. He is, it emerges, having an affair with his nurse; his wife, lonely and abandoned, suffers. Hira's fantasy of a life without loneliness and without responsibility is tied to his own transgressions; he fantasizes a world in which he doesn't have to take responsibility for his actions and tries to realize that world through his creation of the mask. He is both a tyrant and child, searching for a way to excuse himself and manipulating others to try the way out first.

Okuyama willingly plays his part. He's eager to rejoin the world, sure, but he's also eager to expose his wife for her supposed shallowness and lack of commitment to him. What he exposes is precisely the opposite; she knew it was he all along and thought he was doing whatever he could to suture the crisis in their marriage and regain intimacy with her. She struggles early in the film to come to terms with his bandages and the blank face he presents her—emotionally, conversationally, and in the whiteness that swathes his face. She is, in good faith, trying to connect with Okuyama on his terms, which she imagines demand the mask. They had a chance. He ruined it.

It is not only women who have anxieties about how they look.

Okuyama's wife (never named) also wears a mask, according to the film. All women do: in a discussion about makeup as a kind of mask, Hira tells Okuyama that makeup is "also a sign of humility. As long as a woman is a woman, her face isn't worth showing without makeup" (46:53). But women can handle the ability to remove their masks at the end of the day, unlike Okuyama and maybe other men. Makeup is a mask; there is a face beneath it, always. Wearers of makeup are not the living dead. Though the mask of makeup may transform, it doesn't duplicate: makeup does not challenge the 1:1 correlation. The mask of makeup, which requires changing the face only of the wearer (and no one else), does not (anymore) challenge morality as we know it.[39]

Okuyama's wife wears her mask but is a moral and accountable being nonetheless. Okuyama's wife wears her mask and has the unerring ability to identify the mask on others. In fact, Hira's plan was doomed to fail because the masked always recognize masks in others. She knows the masked man is her husband all along. And she isn't the only one: in Okuyama's bolt-hole apartment that he rents for his double identity, the superintendent's simple-minded daughter is not fooled by the mask. She knows masked Okuyama and bandaged Okuyama to be the same man. Women, well schooled in masking, well schooled in loneliness, and well schooled in betrayal, attest to the truth of one of Hira's homilies and lies: people do have a prejudice about faces, but it might be only that—a prejudice. Faces and masks may not be identity. At least, not for those with morality already firmly in place. Which in this film seem to be those who have always lived with masks—the women. (Or—and this is possible given the film's philosophical commitments—the women are offering us a hint at other, better ways of being in the world. Ways of recognizing the other, and our obligation to the other in the face-to-face moment, without relying on essentialized external appearance. This is a deeply Lévinasian moment, in which the other is encountered fully, meaningfully, but without, in Lévinas's formulation, knowing the color of his eyes.[40])

These women are not the doctors (hardly surprising given the historical moment in which the films were shot, and hardly surprising given that so few of them have names, but still worth noting). In *The Face of Another* and *Eyes without a Face* they are the nurses and the assistants: secondary, helpmeets, and in both cases involved to some extent with the doctors, their bosses, lovers, controllers. These are films about relationships and their destruction, films about fathers and husbands betraying their families for the sake of faces. These men are so desperate not to lose face that they lose everything else.

Eyes without a Face plays with this notion of family as the center of its horrific moral compass, parodying in one scene a cozy family dinner with the tyrannical father, the blindly obedient "mother," and the hidden and already dead daughter sitting down together for a lovely meal in that brief moment when she has a functioning face.[41] The mother is also a daughter, owing her face to the doctor-father whose surgical experiments gave her back her life. But Louise may in fact be the actual mother; she is a mysterious entity who, we learn at Christiane's "funeral," appeared in Génessier's life following the death of his wife. Or the "death" of his wife? Is Louise, whom we know had a successful facial reconstruction, the product of another of Génessier's familial experiments, a devoted wife with a new identity? Even as a wife, Louise is, in some ways, a truer daughter to the doctor than Christiane, whose face quickly disintegrates after the meal scene. The life her father gives Christiane is fleeting and she is quickly sent back to her liminal space of in-betweeness; the doctor is a far better father to Louise, whose existence in the world is far more secure. Until one daughter kills another. In *Eyes without a Face*, uncontrolled violence is a feminized activity; the male doctor commits his atrocities in a careful clinical space.[42] Women are perpetrators and victims. And they all have names. This film is a World War II allegory, and, in an ethical recognition, Franju insists that the victims have names.

At the center of this family is Dr. Génessier, patriarch and tyrant.[43] He binds his family to him by ruling them absolutely, drawing both women deeper into his chamber of horrors as they are both party to the acts of violence he performs ostensibly for their (and particularly Christiane's) sake. These are not acts of love (twisted as that may be on its own) but acts of guilt, and anger; his daughter's disfigured face is intolerable, and she may not exist as anything less than perfect. Without a face, she is no different than the dogs on which he experiments; like those dogs, her marred appearance is his fault. And, like those dogs, her anger at her father becomes homicidal. And, yet, Christiane is unable to directly kill her father-doctor, instead using the dogs to execute the act. She does, however, stab Louise at the site of her facial-reconstruction scar, killing her mother-sister and her father's more successful creation. (The Freudian analysis is too obvious to belabor.) With this killing, she begins the murder of her father by destroying his one successful legacy; she frees the dogs to complete the act.

Christiane's actions, however, are not those of rebellion or redemption. They are not motivated by a desire to counteract her father's evil or undo his efforts, nor does she wish to reassert her moral core. Unlike Okuyama's wife, Christiane's morality is long gone. Rather, her anger, her rage, comes from her father's failure to fix his mistakes, his failure to follow through on his promise to give her back her face. She releases his latest potential victim, Paulette, not out of compassion but rather a sense of the futility of the endeavor. Just like all the other attempts, this one would also fail; there is no point in trying again. (Her decision foreshadows the moratorium on organ transplants; despite the potential upsides, it simply was not worth it.)

The living dead have no morality and no identity. However, it was not the accident itself that caused Christiane to become no one. She lost her face, but it was her treatment by others that caused the loss of her life. It was her father's reaction to the event and its repercussions that firmly placed Christiane in limbo. His—and Louise's—insistence on her wearing a mask, for them alone, his manic and desperate attempts to fix her appearance, his collusion in her false funeral, and his tyrannical control over her confinement, combined to steal her selfhood. When she walked away from Louise and her father, she was not reclaiming her life; she had long conceded her liminal state and continued to keep her mask firmly on. She just wanted to occupy her limbo unfettered by those who would continually disappoint her.

Eyes without a Face is concerned with victimhood and patriarchy, but there is no happy feminist ending. For Christiane, there is no (happy) ending at all.

Wilson gets an ending. It isn't a happy one necessarily, but his death releases him from the possibility of endless limbo and liminality occupied by Christiane. That limbo is a very real threat, presented in the form of the overflowing Company waiting room filled by men waiting, waiting for their next rebirth. Wilson does get reborn, in a kind of way, not as a new identity, but as a once again moral being. His death (not to glorify murder too much) is a release and a return, making true the funeral of Hamilton and making his fractured identity whole.

Hamilton/Wilson is not, as I've said already, a hero. He abandons his family twice over: for his work, early on, and for his rebirth, later. The film critiques Wilson's quest to be a lone masculine hero, a 1960s ethos represented by Davalo, the post-op therapist.[44] He explains Wilson's status to him in glowing terms: "You are a bachelor. You are alone in the world, absolved all responsibility except to your own interest. Isn't that marvelous?" It's not marvelous, and Wilson is not marvelous. He allows his wife to mourn him, and she does. Despite the distance between them, she is, as the film is careful to show us in a scene with her and Wilson, bereft by his loss, by the breakdown of her family. Wilson spends much of the film trying to rebuild a family while at the same time isolating himself; he rejects attempts to create community with his neighbors, but quickly finds himself with a girlfriend. Both, however, are false relationships: neighbors and girlfriend were sent by the Company to monitor him. He is in fact utterly alone; the closest he comes to a family is the Company itself. After all, they gave birth to him.

And they kill him. It's time to talk about the deaths.

THE DEATHS

One of the central premises of this chapter is that the deaths in all of these films constitute a kind of punishment for violating the invisible boundary between acceptable and unacceptable self-improvement and change. Unacceptability here is partly about erasure of self and attempts to eradicate time; trying to ignore the effects of experience and the dictates of history dooms one not only to iterative failure but also to change in unpredictable and monstrous ways. The postwar context heightens the stakes of recognizing and honoring the past rather than trying to replace it. As I have shown in the previous chapter and will explore again, this is something that has changed; we now accept attempts to recapture what once was. At the same time, I've argued that death serves as a kind of release from the limbo hell of eternal facelessness; better to be dead than the living dead.[45] These two notions seem to have a contradiction in principle: how can punishment also be release, or, how can the denial of punishment be a punishment in itself? Setting aside the obvious response that some people long for punishment, and that for some punishment *is* release, I'll focus in particular on the cases outlined here.[46] In order to do so, it is important to consider exactly who dies, who kills, and who, in these instances, is being punished and for what.

Seconds is the most straightforward instantiation of the punishment motif: Hamilton becomes Wilson, embarks on an epic journey that goes nowhere, fails at his transformation, fails at his attempt to retransform, and is killed at the hands of the institution that enabled the whole process. Hamilton's death (as Wilson) is the endpoint of a hamster on a wheel, running fast and staying in the same place. Eventually, the only place he has to go is away. He's out of other options. His grand plans of an epic journey, like so many in the films of the decade, end nowhere.[47]

And, certainly, Wilson's death makes true the lie of the corpse of Hamilton, bringing him back to his boring suburban life and identity and, in some ways, erasing everything that happened in between. Except that the death itself speaks to the kind of self-awareness and growth that the "journey to nowhere" rhetoric denies. If Wilson-as-Wilson truly learned nothing and went nowhere different than Hamilton, he wouldn't have actually died but rather would have continued his life in the endless waiting limbo or offered the name of another client/victim to suffer the same fate. His refusal to do so demonstrates personal development; Hamilton was so unable to see beyond his personal goals that he destroyed his marriage and all other connections and relationships. Wilson, partly through his visit to his/Hamilton's widow and partly through his experiences in his new life, has a more global perspective and a greater sense of empathy and responsibility to others. This sense of responsibility leads to his death, his unwilling death (let's just call it murder) but not his meaningless death. Wilson's murder does work in the sense of a classical tragedy, in which the hero struggles and fails to overcome overwhelming forces, but does so in a noble fashion that grants him moral victory. Wilson certainly gains the moral victory against the company, and his death serves to correct his earlier deceptions by making them true. For Wilson, death means recalibrating, following an aberration from the natural order, tragic (to him) as it may be.

Is it a punishment, and, if so, for what? From the perspective of the Company (said overwhelming force), it certainly is a punishment, both for Wilson's failure to adjust to his new life and his refusal to offer a new body to replace the one initially wasted on Hamilton. It's not a punishment (from their perspective) for attempting to change too much but, rather, for not changing enough. Except in the most important of ways, of course: Wilson did change enough to not commit one of the same mistakes twice. This time around, he put the needs of others before his own. However, his eagerness, or at least his desperation, to be reborn again shows only limited growth; he understands that it is an enterprise doomed to failure, but he wishes to fail again.

Death puts a stop to this endless cycle, this limbo of ever-changing faces and bodies that is a stationary journey. It's a punishment but also a release. It's an unasked-for consequence—a terrible one but also the only one. The Company punishes Wilson but redeems Hamilton at the same time. So, yes, Wilson is being punished for trying to change too much, or too many times, but the nature of the attempted change is one that saves the formerly doomed Hamilton. Wilson is punished, but Wilson is no one and never was anyone. Hamilton was punished by becoming Wilson, and Wilson's existence holds Hamilton's hostage. The death of the latter releases the former, and we (the audience) are given license to mourn rather than celebrate that death, hate rather than celebrate the Company.

Death is far from romanticized, however; in this intensely corporeal, sensual film, obsessed with sweating bodies and their consumption—of food, sex, alcohol—death is equally physical. Much like the sign advertising used meat on a truck that we see in the film, Hamilton too becomes seconds, nothing more than a slab on an operating table to be used for someone else's consumption. It's shocking in its mundanity, and it's shocking in another way: so rarely does the Hollywood hero die so unglamorously.[48]

The Face of Another plays with the question of punishment by death. Okuyama is the obvious target, living as he does in faceless limbo and challenging the natural order of identity through his embrace of someone else's face, but it is not he who dies. Instead, he is the murderer, killing the doctor who enabled his (reversible) transformation in a seemingly direct act of either retribution or insanity or both. One narrative would frame the attack as the inevitable result of Okuyama's dangerously fractured identity, accurately predicted by Hira himself. Indeed, Hira foreshadows his own death, wondering which of the three characters engaged in the seduction deception (Okuyama, his masked self, and his wife) would ultimately snap and commit murder. Hira's prescience is only partial, however; he never dreams that he would be the actual target of the killing.

He should have figured it out. Okuyama, abandoned, alone, fully aware of the consequences of his actions and his own missteps, is filled with rage. Some of that is directed toward his wife, some toward himself, and some toward his mask. Like his identity, Okuyama's rage may have remained fractured and possibly in check, or at least nonhomicidal, satisfied instead by acts of assault such as the one he committed on an innocent female passerby. But

Hira heightens Okuyama's rage, pushing it to the breaking point by putting words to Okuyama's ineffectualness, telling the police officers that he was but a harmless mental patient who escaped custody. Already struggling with the loss of his lover due to his inability to consummate the relationship without a mask, Okuyama is faced even more starkly with evidence of his unmanning. Hira puts words and reality to Okuyama's ever-growing physical and emotional impotence, loss of manliness, and disempowerment that began with his loss of face but culminated in his refusal to accept being in the world with said loss. (And, yes, the film blames Okuyama. And Hira. And it also blames the world for being an inhospitable place.) Okuyama's affair with his own wife is the physical embodiment of his failures; Hira's insistence on Okuyama's docility, his claim that Okuyama poses no threat or danger, is the social confirmation of Okuyama's uselessness. So Okuyama claims back his identity and virility by striking at the embodiment of its loss. He kills Hira and regains himself.

This reading implies that Okuyama's decline is due not to his loss of face per se but to his own failings in responding to this tragedy. Hira, acting in direct opposition to his professional mandate to help Okuyama overcome his initial reaction, validates and even heightens Okuyama's sense of impotence and powerlessness without a face. Rather than offering coping mechanisms for this lack of face, Hira offers a new face. So Hira is punished for creating the conditions of possibility for too much change, while at the same not enabling any real change at all. Once again, the true change is that which brings about the murderous moment: Okuyama realizes that he needs to reclaim his masculinity, his power, his identity, his individuality. Living life as an anonymous man in a mask only increases his incapacity. Okuyama insists on the danger that he poses—and then demonstrates it—as a direct counterexample to Hira's fantasy of a world of no identification, no accountability, and no morality. Okuyama knows that what he is doing is wrong. He does it precisely because it is wrong; a world of masks is *not* a world without crime.

To underscore this point, Okuyama stabs Hira in a sea of faces, or rather facelessness; the crowd that surrounds them is populated by blank heads.

Okuyama, like Wilson, is not a hero. He's a deeply flawed and often cruel man, and probably always was. He is a killer, and he is probably wrong in insisting that he is not at least somewhat mentally imbalanced. He is in thrall to his mask, and his mask exacerbates the worst in him. It also changes him. Faces are destiny, at least a little. But they also, as highlighted by Okuyama's wife, aren't the only factor in identity. People can work within their masks, or their faces, or their lack of faces, to be people they ought to be and wish to be. Or they can give in to the perceptions of others, their own inclinations, the seductive pull of ideal appearance. Okuyama does, and hates himself for it. He also hates Hira for it, and he kills Hira for it.

More people die in *Eyes without a Face*. Many more. This film, alone of the three, has innocent victims, namely the young women whose faces were stolen to try to heal Christiane. The medical logic of the film allows them to live without faces, or at least to survive their face-removal surgeries, but the larger logic of the narrative requires that facelessness renders them half-alive at best, a condition to which at least one responds by leaping to her death. That scene is ambiguous; Edna is leaping away from the house, away from captivity, away from the horror. She is escaping, but the only real escape available for her is death. We don't know the story behind the first corpse we encounter, the one that provides the body in Christiane's false funeral, but we do know that she was also lured by Louise to Génessier, and we do know (based on his ability to falsely identify her) that she has no face.

There are also uninnocent victims. Christiane is an uninnocent victim. Initially a car accident survivor (sort of), her innocent state is increasingly tarnished by her collaboration (a poignant word in post-Vichy France, obviously). But her fate will not mirror those of the innocent victims. Her punishment is not death, not yet. Instead, she has to emerge from the suffocating shelter of her prison and face the world. Still in limbo, still masked, Christiane has to find out what it means to be faceless among those who do not have as their mission restoring her appearance. Christiane has to find out what it means to be seen by others. But she leaves with her mask on.

In Arendt's terms, Christiane denies her place as a member of society, and shirks her obligations. In Richard Sennett's terms, by contrast, she is always only a public character, with no possibility of intimacy or relationships, always hiding her private self. Even at home, she is in public, and when in public she is hiding behind her mask and all it represents. She is amoral and emotionally silent, not because of her mask but with its assistance. Christiane has the chance to change this as she walks away, but she chooses to leave her mask on and be in public—and be hiding in public—going forward. She leaves her house but carries her prison with her. It isn't her facelessness but her relationship to it that keeps her isolated. Sennett has argued that the nineteenth century saw the dawn of silence in public in contrast to private intimacy; Christiane shows the ways that the public becomes the private, and intimacy becomes impossible.[49]

And there are guilty victims. This is the easy part, the easiest part. Both Louise and Génessier deserve to die, and they are killed in ways that mirror their crimes. Louise stole faces and identities, which is cruel and horrific in its own right, but that isn't the crime for which she is punished by death. Stabbed at the site of her facial reconstruction scar, Louise dies for her grand deception in changing her face too much. That scar, hidden by her ever-present pearl choker, is the mark of her denied limbo, her manipulation of what should have been a half-life of averted eyes in a half-face. She dies for gaining back her life, for having eyes *within* a face, for managing what Christiane does not. Christiane murders her not out of revulsion for Louise's kidnappings and blind obedience to Génessier, but for her failure to act as a true twin and double, with a shattered face mirroring Christiane's own. Louise is guilty of many crimes, but she dies for the one that highlights Christiane's ethical perversion rather than her own. There is a morality in *Eyes without a Face*. It's twisted and fractured and very, very ugly, but it has a code. Génessier has a code of obligation, and Louise has a code of reciprocity. It might be Christiane who is truly amoral and self-serving, and, so, we do not pity her, despite the horror of her condition.

Génessier is the guiltiest victim. His cruelty is unspeakable, the horror he inflicts is torturous. It's clear, it's not up for debate, it's one of the central focuses of the film. But that's not why he dies. He is killed (by the dogs he tortured, released by Christiane) for his failure to make good on the promise of the faces he steals. He is killed not for maiming young women and driving them to their deaths but for not succeeding in redeeming Christiane's face. To add insult to (literal) injury, he restores Louise, Christiane's rival and double. Christiane may be angry at her father, disgusted by him and appalled by his actions, but she is driven to murder because she has lost faith in his ability. She can live with him as an evil doctor but not as a bad (or at least a not-good-enough) one. Because he failed to restore Christiane to life, he dies. His biggest sin, in the logic of the film, is promising too much.

In none of the films then, does anyone die for trying to change too much, or for violating the limits of acceptable self-change—at least not exactly. But, in each case, someone is murdered for failing to accept the conditions of facelessness, failing to adhere to its implications and ramifications. So people die for violating the destiny of the face and appearance. People die for trying to rewrite (or even eradicate) history through rewriting (or eradicating) its effects on the face. Faces matter because history and experience matter. The accuracy of the face in representing its history and experiences matter.

REWRITING HISTORY

These films were all made at a time when there was considerable interest in the possibilities of rewriting history, or at least reimagining a future apart from very particular histories. The contextual issues aren't very complicated to tease out: we're looking at post–World War II Japan, France, and the United States. One loser, one conquered country, and one victor, each struggling with the effects of the war and its own identity and sense of self. Japan, a culture obsessed with pride and self-containment, lost face on the global stage; *The Face of Another* imagines a different kind of losing face and its meaning and implications in the wake of a tragedy caused by science and knowledge. France, a country that imagined itself resistant, senses whisperings of collaboration, a chorus that grows stronger even as it is ignored or reframed. *Eyes without a Face* thinks about the moral effects of collaboration even for the seemingly innocent or accidentally implicated. And *Seconds*, a film based in an ever more prosperous and modern United States, thinks about the real meaning, or meaninglessness and lack of freedom, of all this prosperity and modernity in the post-McCarthy moment. The two films from losing countries are particularly invested in cultural honor and collective notions of identity, while the American film plays with notions of radical masculine individuality and its entrenchment. What is especially intriguing is that, despite these differences in focus, all three films ask questions about progress, identity, and change, and they answer them through stories of facial manipulation, modification, and substitution. They all do it through radical interventions with, and replacements of, the image that identity presents to the world. They all do it through the face.

Or, rather, they all do it through faces. There is no one face in any of these films; each is interested in doubling, in masking, in dual identities. Even as each shows that in the end history cannot be rewritten, and the future cannot be reimagined without taking into account the lessons and limitations of the past, they all offer a kind of way out of the stability and stagnation of the present. They all suggest that there are multiple versions of the present through multiple versions of the self. Everyone has a double. But, in all of these films, the double is worse than the original self, even as the original self has changed. Even as the original self, through attempts to alter the past, has become monstrous. These films absolutely reject the makeover narrative of locating your true self through external manipulation; the true self is the one that people have become over time, not the one they must change to be.[50] The take home message is that we'd best work with what we have rather than try to replace it entirely.

DOUBLING

Seconds tries to replace the current reality entirely. Hamilton becomes Wilson becomes another failed life becomes dead. Wilson is a way out but a bad one, a double that is attractive on the surface but without the ties and relationships whose value Hamilton failed to recognize until it was too late. What Wilson offers Hamilton is the opportunity to appreciate what Hamilton always had; what Wilson offers Hamilton is the knowledge that Hamilton's life was better. Wilson shows us that Hamilton didn't need radical transformation from the outside. What he needed was modification on the inside. The outside, in the end, was just fine.

Eyes without a Face thinks about doubling in a less individualistic way. Christiane is infinitely reproducible by or reducible to a series of young, vaguely similarly featured women. Any of them serve the purpose of helping Christiane become, once again, herself, a point underscored by the final potential victim and police plant, Paulette Méredon. Paulette is instructed to dye her hair a lighter color, which she objects wouldn't look right. Her objections are set aside because they don't really matter. She just needs to have her coloring sort of resemble Christiane's to be an acceptable potential face donor. Or another way to put it: all these young women are basically the same, even Christiane. Individual identity isn't the point. Even Christiane's individual identity isn't the point, and her father's obsessive commitment to restoring it by restoring her face is part of his moral failing. But Christiane ignores all these women (until she doesn't), ignores their innocence and their suffering. Once their faces are removed, she doesn't see them at all; before their faces are removed, she looks and sees only herself. They are doubles who show her nothing because she is too narcissistic to see anything but her idealized reflection.

Christiane has another double as well. Louise serves as her counterpoint, the one whose face *was* successfully restored, the Frankensteinian daughter who isn't a monster. At least not visually. But of course Louise *is* a monster, and her new face is part of the reason why. She might not have looked like one, and that too was part of the problem and maybe one of its causes. Louise shows Christiane what not to do, what not to wish for, what not to become. Christiane either doesn't care, or doesn't pay attention.

The Face of Another is both literal and abstract in its presentation of doubles. Okuyama becomes a double himself when he dons the mask that faithfully reproduces its model, another person entirely. But Okuyama is also doubled by the facially marred young woman who kills herself because of her loneliness. She represents one path that Okuyama could have taken; rather than killing another, he could have killed himself. She also shows that Okuyama's suspicions about taking off his bandages and exposing himself to the world are quite correct: the world would reject him, and he would be alone. But Okuyama is also quite wrong in thinking that he needs to remove his bandages for that fate: he has, in fact, already rejected the world before it could do the same to him.

Okuyama has a third double in Hira, the would-be faceless man who longs for the anonymity that he has provided for Okuyama. Hira's extramarital affair with his nurse tracks Okuyama's attempts to trap his wife into infidelity; Hira's loneliness and cruelty foreshadow the effects of marital deception and destruction that will befall Okuyama. A fourth doubling exists between Okuyama's wife and the superintendent's simple daughter; they are the only ones who can see behind Okuyama's mask of another to accurately identify him, a skill that even he finds elusive sometimes. Because he has lost himself.

So many doubles, all serving to (ironically) underscore the nature of individuality, the importance of distinctions between people, the ways that appearances do not tell you everything about character but actually do offer quite useful ways to identify specific figures. The doubling motif is particularly poignant in films about faces, facelessness, and attempts to reverse the latter condition with the help of the former. Films about faces care about distinguishing one person from another, even as they play with drawing parallels between characters.

CONCLUSION: STAKES FOR THE FACE

What we have in this chapter are three films that all grapple with postwar modernity and technology, that all challenge the role of the doctor and the ethical value of medical advances, and that all engage with the destruction of the family. They all have central doubling motifs, and they all draw our attention to the links between appearance, identity, and identification. In all of them, people die as a direct result of breaking those links. All the films have murders, murderers, and victims, innocent and otherwise. And all the films get to all of these issues with their obsessive interest in the face, in its manipulation, transplantation, and replacement, in its alteration as a way to rewrite history and reimagine the future, and the failure to do so. These are all films that tell horror stories with the face. Why?

Well, why not? Faces are pretty damn important. So let's be more specific: why does the manipulation of faces feature so centrally in horror films? And why, really, do so many people have to die? Actually not *why*—there is death because these are horror films. So, to be yet more specific, why does facial manipulation in particular lend itself so easily to the horror and murder genre? That's a better question.

Answer: Because these are all films that cut to the core of what it means to be human and what it means to be inhuman, both individually and in relationships between people. And it turns out that cutting to the core of the face is the best way to cut to this core. Because faces are the way that we understand humanity, for better and (here) for worse. If we change faces (too much or too obviously—this line still exists), we make it impossible for the rules of humanity to apply. Hira was right: without faces, we would have (and we would need) a new morality. Hira was positing a kind of moral anarchy, and there he got it wrong, but he was correct in that the rules of interpersonal relationships as they are currently configured depend on the face-to-face interaction, and without that we would need new rules, ones that we have not yet figured out. The women in the film might have figured them out, but they aren't sharing their method. So that's why facelessness is not just a problem but also a fundamental challenge to the social order.

The move these films make is to equate taking on the faces of others with facelessness. The challenges to the social order are posed *not* when people lose their faces. It's not the bandages, the disfigurement and deformities, the injuries and their treatment that, in and of themselves, are the punishable offense. Sure, they are a problem—they are hard to watch, and they dehumanize both the victims and those around them who don't know what to do. But they—we—don't know what to do because the usual rules of social interaction and morality still apply, and the bandages and the disfigurement make it hard to apply them, to look people in the face and to identify them by what we imagine to be their injured and deformed selves and self-images. The bigger problem is when we don't know to stop and ask those questions because the faces we are seeing are not truly their own, aren't truly anyone's. That is true facelessness.

So Hira, in a way, serves as the proxy for Lévinas, whose ethical system requires seeing the face of the other in radical interrelationality. (We'll talk more about Lévinas in chapter 6.) Sometimes people have no faces, and we deal with that. But if no one had a face, no one at all, well, then we would have to figure something else out.

But in these three films, most everyone else does have a face, or they do until people contrive to remove them. And that's the punishable offense, and people get punished for trying to appear and, in some cases, to be other than what they are. It all comes back to cheating in the end. Or, to be more exact, since most of us cheat a little or even a lot with our faces, it all comes back to cheating too much. Radical transformation may be possible, but at a very high price; if you try to change yourself too much, you will be punished. Maybe by and with your own monstrosity, which mirrors the monstrosity of your enablers. This is, in its way, a fitting punishment for taking the face of another without consent. For attempting to replace rather than recuperate, to erase rather than engage. This is the (not so hidden) message of numerous narratives of self-discovery; what is ultimately discovered is that you should not, and indeed cannot, mess too much with nature or especially with time and experience. There is a line, and when it is crossed, you may well be crossed off. It's terrible, of course, to steal facial flesh off a living donor, and it's wrong to abandon your family, but what's really bad is trying to be who you are not. It's bad, and it's bound to fail.

These ideas didn't go away after the 1960s. The line around what constitutes too much change may be a fluid or movable one (with, for example, our growing tolerance for the erasure of time and the recapturing or highlighting the hidden self), but it is an extant one. Many of the contemporary debates around facial allografts that I explore in the next chapter resonate with these earlier and sometimes catastrophic representations. The bioethical and critical discussions leading up to the first face-transplant surgery were obsessed with the idea of going too far, fearful of the possibilities and implications of the procedure. Some critics raised concerns about profit, as explored in the movies. There were debates about the need for the surgery itself, not just given its risks but also its stakes, namely identity and the nature of personal experience. Or, to put it another way: maybe the ennoblement of suffering outweighs the benefits of facelessness.[51] Some were suspicious of our powerlessness in the face of progress and what that might mean. The question of consent, and its impossibility in such desperation, was repeatedly raised. And all wondered if the results of the surgery were worth the (immediate and ongoing) risks. These films explore not just what it means to get a new face but also what it might mean to live without one. They are all critical of the patients and recipients (the faceless, or formerly faceless), exhibiting the same inability to make sense of the quality-of-life concerns that the bioethicists exhibited so many years later in debating the now very real possibility of face transplants—or, these concerns were dwarfed by other, greater fears.

But the greatest fear, or greatest question, of all is one that still lingers: what does an interpersonal ethics look like without a face-to-face relationship? Or, when the actual face no longer provides a singular or unique index to a specific person, how do we relate to others?

These are not just questions of facelessness, or of the digital, or of transplantation. They are, as Arendt, Sennet, and Goffman teach us, long-standing concerns. But they are newly urgent in this moment. That is the real source of the fear. Later in this book, I explore what a new, nonfacially oriented relationality might look like. It's already here: the world is full of avatars who provide multiple indices to the specific self. And, really, it's always been here. The 1:1 index has always been vexed. We have always been performing and mediating ourselves. Changing the face does not affect that.

But first we must track what changing the face actually does do. And what it did do. Enter Isabelle Dinoire.

CHAPTER 4

Decoding the Face-Transplant Debates

INTRODUCTION

When the first (partial) face transplant happened in France in 2005, it shocked the world. In retrospect, it's hard to remember exactly why. It's hard to recall the motivations behind the discomfort, the disgust, and (this is a strong word but an accurate one) the fear. But the reactions didn't occur in a vacuum, and they weren't unprompted. They were part of a larger conversation rooted in the news media and emerging directly out of the bioethical and surgical debates preceding the surgery. The earliest news narratives were cautiously positive, celebrating the practical advantages of the procedure for (recipient) Isabelle Dinoire and future patients. As details about the specific case emerged, coverage shifted dramatically in tone as both the doctors involved and Dinoire herself were critiqued and criticized. While broad bioethical concerns provided the ostensible framework for these challenges, the underlying narrative rested on a discomfort with Dinoire's own right to the procedure, which occupied an ambiguous space between medically necessary reconstruction and elective aesthetic surgery. But there was a lot of room in the space between, and it filled—really rather quickly—with critique and disgust. It could, as I show, have gone another way.

The world's first partial facial-allograft surgery took place on November 27, 2005 at the Centre Hospitalier Universitaire Nord in Amiens, France.[1] The face of now deceased recipient Isabelle Dinoire had been mauled by her dog after she passed out from an overdose of sleeping pills. The lower half of Dinoire's face, including her nose, lips, and chin, suffered damage so severe that she was unable to eat, receiving nourishment from a feeding tube in her stomach. Following admission to the hospital, doctors ruled routine reconstruction impossible, suggesting instead a radical and heretofore never attempted transplant intervention. Hospital specialists, led by maxillofacial surgeon Bernard Devauchelle and surgeon Sylvie Testelin, quickly assembled a team of specialists, including hand-transplant pioneer Jean-Michel Dubernard. No stranger to controversial procedures, Dubernard had weathered tremendous criticism for rushing in to perform the first hand transplant in 1998, which was ultimately removed at the request of the recipient. For Dubernard, Isabelle's situation presented a pressing and clear need to action that was immediately obvious: "Once I had seen Isabelle's disfigured face, no more needed to be said. I was convinced something had to be done for this patient."[2] To date, with the exception of the Cleveland Clinic, which received IRB approval for facial allografts in 2004, hospital ethics committees and IRBs had consistently denied permission for the surgery, despite robust experimental and clinical research on the viability of the procedure. But Isabelle's case, the team felt, was strong enough to challenge the status quo, offering enough medical and psychological benefits to outweigh the conventionally stated risks. After four months of extensive medical, scientific, and psychological testing, including MRIs to ensure the functionality of the part of the brain that controls the (missing) lips, the hospital ethics committee granted the team permission to proceed. All that was missing was the donor.[3]

Two months after the approval (and six months into Dinoire's hospital stay), an appropriate (ish; the suicidal conditions of her death were suppressed) donor was found, launching the transplant team into action. The surgery itself was highly intricate, involving microthreads to attach the major arteries, which had to line up exactly between the donor and the recipient. The initial key to success was the flow of blood between Dinoire's veins and the donor face, allowing the donor flesh to maintain life and viability. When the arteries were unclamped at the end of the fifteen-hour operation, the lips and skin pinked up, marking alignment and early victory. However, the journey was far from over for Dinoire and her doctors, both medically and in the theater of public opinion.

The major risks posed by the surgery extended far beyond the danger of fifteen hours under anesthesia while undergoing a major bodily procedure. On the timescale of risk for Dinoire, fifteen hours was but a blink of the eye. Following the surgery, she would need to be on immunosuppressants, likely for the rest of her life. Should she stop the regime, her face would literally slough off. On the regime, her entire body would be vulnerable to a host of illnesses against which a robust immune system normally protects. Transplant patients are rigorously evaluated for their fitness to adhere to this lifelong medication regime and its associated vulnerabilities.[4] But in the case of the facial-transplant recipient, the immunosuppressants are but the tip of the psychological-challenge iceberg. Far more uncertain, with almost no precedent to date (beyond some similarities with hand-transplant recipients), was the question of what Dinoire would see every time she looked in the mirror, namely, the face of another. There was no proven psychological-fitness test to evaluate how she (or anyone) would handle that.[5]

Was the surgery—which was not, strictly speaking, lifesaving—worth the risk?

At that point, almost all ethics committees and IRBs said no, despite lobbying attempts and applications from surgeons at a variety of institutions across the world.[6] The popular press wasn't so sure, at first. With little precedent for the procedure itself, reporters embraced the ambiguity around the classification of the surgery and borrowed from both transplant-surgery narratives and cosmetic-surgery narratives. They acknowledged the quality-of-life issues at stake and the very real changes the surgery could make in Dinoire's life. At the same time, they were concerned about the implications for high-risk facial manipulation. It was the latter narrative, undergirded by a broader suspicion of cosmetic interventions, that quickly came to dominate. But, as I will show, there was some initial space, largely forgotten in the later media maelstrom, for a different response.

RISKS AND BENEFITS

The emergent suspicion of the surgery was fed by a variety of concerns over facial manipulation and its implications for identity construction, but it rested on a stark bioethical calculation of risks and benefits. There were the obvious risks of the surgery itself, but these were dwarfed by the long-term implications of an immunosuppressant drug regime required to protect against rejection of the facial tissue. With traditional aesthetic surgery, there is no issue of rejection, and thus no lifelong regime of immunosuppressants, which both require compliance and significantly weaken the immune system, placing recipients in positions of increased vulnerability to a variety of infections and illnesses. Aesthetic surgeries may, like facial allografts, be elective (in the sense of not strictly lifesaving) from a technical perspective, and may indeed be far less therapeutic in nature than the face transplant, but they also represent significantly less long-term risk to the patient. Not so with a variety of organ-transplant surgeries, including heart, lungs, and liver. Kidney transplantation is a more complicated (and thus highly relevant) case because patients technically can live as long-term dialysis users for up to ten years, though there is significant mortality among these users over that time.[7] Without dialysis those with renal disease would die, but with dialysis they can survive for a while without transplants. All things being equal, antirejection medication would make a well person sick. But well people do not get transplants. Not normally.

Given the benefits offered by the transplants, the risks assumed by a compromised immune system and the requirement to adhere to a lifelong regime do not seem to present a complicated calculus of necessity. Which isn't to say that, given the scarcity of vital organs in certain countries, there isn't a very careful screening and assessment protocol to determine who receives these life-saving organs and who—because of psychological unfitness, bodily abuse, age, unlikelihood of surviving with the organ, or a variety of other reasons—is condemned to die.[8]

Dinoire and subsequent potential recipients of FAT were not in immediate life-threatening danger. While the quality-of-life implications for the transplant were clear, bioethicists struggled to figure out how to balance these against the risks, particularly given the individual nature of each recipient's experience of facelessness. As Taylor-Alexander has outlined, surgeons carefully framed the surgery in terms of its ethical advantages; I emphasize their focus on patient need and the stakes for facelessness.[9] As the subsequent bioethical and journalistic debates make clear, this strategy was ineffective at best. Equally at issue was the groundbreaking nature of the procedure; while the technicalities were well tested in hand-transplant procedures, there was little precedent for the psychological outcomes. And, like all transplant operations, the risk of rejection was present and, in this case, particularly problematic; in order to prepare the damaged face for transplant, what little tissue that remained had to be removed. Should the face be rejected, the patient would be worse off than before.[10] The relative weight of these issues was in flux even as the first surgery was conducted.

The flux was a challenge for reporters struggling to get a handle on this unprecedented story. In early articles, journalists drew on these critical bioethical responses as part of their framework, but they also accessed quality-of-life arguments as the balance between these pieces was negotiated. As my survey of English-language newspapers of record in England, the United States, and Australia shows, the ambiguity—around the nature of the surgery itself and its implications for the bioethical argument and the journalistic framing—was quickly resolved.

The resistance of the procedure to categorization—neither exclusively cosmetic nor lifesavingly therapeutic—presented a framing challenge for those covering and discussing the surgery. The initial coverage spoke to both aspects of the surgery by focusing on the medical breakthrough and the quality-of-life advantages it conferred. The next wave of responses took on a highly moralistic and very specific tone, reading the surgery in the context of Dinoire and her doctors and shifting the debates from general bioethics to the specifics of Dinoire's life story. As ethics entered into the debate, the medical aspects of the surgery were eclipsed by personal details, which scripted the risk-benefit analysis and reframed it in terms of psychological fortitude rather than therapeutic benefit. The ambiguous categorization of the procedure left room for a reading of the surgery as unnecessary, and of Dinoire as an unviable candidate and exemplar of the dangers of unchecked medical advancement. As an attempted-suicide survivor, however, she was presented not as a victim of her doctors' bad judgment but as the cause of her problems that she tried to fix by being a coconspirator in what reporters soon dubbed "the face race." If the doctors were competing to perform the first transplant, Dinoire was, according to this narrative, a willing and eager participant.

Journalists were not the only ones eager to fit the surgery into neat categories. Both the bioethicists and the surgeons slotted the procedure into preexisting frameworks for their own strategic and rhetorical ends. But they had different rhetorical (and practical) goals, and so they used different frameworks to achieve them; the bioethicists largely opposed the surgery, and the surgeons largely supported it. They generally took opposite perspectives, but they agreed on one vital thing: the unique status of the face and the importance of thinking about its emotional resonances. Both approaches, as we'll see, engaged in what I call psychology talk, an informal way of thinking about the abstract emotional implications of facial disfigurement and attempts to treat it. They did it in opposite ways. Surgeons argued for the psychological benefits and indeed necessity of the face transplant, where bioethicists pointed to the potential psychological challenges that accompany a new face. For both sides, the patient's psychology was vital; they disagreed on how that played out in real life. They also disagreed on how that played out in real time: surgeons focused on the stakes of psychological unhappiness in the patient's present condition of facelessness. Bioethicists worried about the psychological unhappiness that would come in the future. So the bioethicists presented the procedure as a (highly and even unsupportedly) risky cosmetic surgery, and the surgeons considered it a lifesaving therapeutic endeavor. The difference lay in the imagined psychological stakes. The centrality of psychology talk emphasizes the particular challenges and unique aspects of talking about the face. There is less psychology talk in whole-organ transplants. Far less.

In the first half of this chapter, I chart the coverage of the surgery in major English-language newspapers during the week following the announcement of the surgery. I analyze the emergence of a highly critical frame in the news, arguing that the mounting condemnation developed as a way to deal with the ambiguity of the procedure and render it cosmetic, elective, and, in bioethical terms, not worth the risk. The debate is highly specific, focusing on the particular instance of Dinoire's surgery rather than the procedure in general. While there were some broad questions about face transplantation as a practice, the focus of the backlash was on the specific characters involved in Dinoire's surgery and, indeed, Dinoire herself. The journalistic presentation was not an abstract discussion about bioethics in general; it was grounded in the details of the case, and the debate, like the case itself, quickly resolved into extremes. The initial coverage showed willingness to accept an expansive notion of what constitutes medical practice, a possibility opened up by the quality-of-life implications for the surgery. But that same ethical framework quickly polarized the debate, reinscribing the divide between lifesaving therapeutic interventions and elective and therefore optional cosmetic procedures.

The details of the operation seemed tailor-made for media attention, with its controversial medical team, its suicidal first patient and (as later revealed) donor, and its pathbreaking precedent. It was, in a time before we used the word regularly, clickbait, the perfect storm to incite massive attention and discussion.[11] The PR for this first surgery was badly handled; the timing was perhaps precipitate, and the procedure itself was always going to be evocative and scary, and not just because of what bioethicist Arthur Caplan has described as "the yuck factor."[12] It's a technical term—really—which he defines as an instinctive response of disgust to new technologies. The face transplant is definitely, in this sense, yucky. Sara Ahmed argues that we police the boundaries of the body as a way to differentiate, to offer a logic of hierarchy and exclusion; meddling with the skin meddles with these mechanisms and with our own structural sense of order.[13] These theories explain, partly, the overwhelming professional and popular media attention paid to the procedure. But it was much more than that. At the heart of these debates was a growing awareness of a new kind of bioethical problem, and an emerging need for a new set of relational ethics. This kind of transplant involved not a new kind of physiological procedure but a particular set of unknowns. The medicine of the surgery was by this point well rehearsed; the psychology of it remained entirely unprecedented. The physiology of the face and the facial transplant is stable; the psychology of the face, well, that is an open question that this surgery might have begun to answer. And, in so doing, it might have told us something we didn't know about identity, and something we hadn't yet figured out about relational ethics in the absence of a stable face. But people were very nervous about the answers. So the debates raged.

PSYCHOLOGY TALK

We've already explored the intimate relationship between transplants and bioethics. We've considered the difficult challenges these surgeries posed in terms of limited availability (who should live and who should die, and who should decide) and the risks of both the surgeries and the immunosuppressant regimes they subsequently require. We've seen how these challenges were navigated for whole-organ transplants and the new pressures raised by reconstructive hand surgery, responses to which are still evolving.

While my stance is critical of the starting assumptions embedded in the bioethical argument that the surgery, because not strictly livesaving, is cosmetic and also not really therapeutic, I want to cut the discussants a little slack. As we've seen, when it came to the particular constellation of issues around facial allografts, there wasn't a whole lot of bioethical precedent. Whole-organ transplants aren't exactly the same; people die in pretty short order without them. Cosmetic and reconstructive surgery isn't exactly the same; people don't have to take immunosuppressant medications following those procedures. Hand transplants aren't exactly the same. They should be, or they should be similar enough, dealing as both do with visible foreign grafts that often require tissue removal before the surgeries and risky medication regimes afterward. They both potentially evoke strong psychological reactions. Except that one deals with the face. And the face is a special case. We may know the backs of our own hands and be shocked to see the backs of someone else's at the end of our arms. But, as bioethicist Françoise Baylis wrote in a discussion about face transplants, "clearly the hand is not intimately linked with personal identity in the way the face is." True to bioethical form, Baylis injects the temporal element, raising concerns about the way the transplant would affect future relationships and the resulting effects on individual well-being: "In all likelihood facial transplantation would profoundly affect both intimate and forlorn spaces between self and others, which in turn would affect identity formation in potentially disruptive ways."[14] Is that effect and all it might entail, coupled with the vulnerabilities imposed by immunosuppressants, enough to justify denying the face-transplant surgery on ethical grounds, even as the hand transplant was growing in regularity? Many, many bioethicists thought so. Because they thought the greater psychological risk was in the future rather than the present. Because they thought the surgery was cosmetic, and they thought that cosmetics were not reason enough.

It might be jarring to group those undergoing elective plastic surgery with face-transplant recipients, but there is a real continuum, especially if we take seriously the cultural models imposing bodily norms that we explored in chapter 2. And that's exactly what bioethicists did by thinking of the surgery in terms of a risk-benefit analysis that doesn't take into account quality of life. Many called for the patients to bear the brunt of the social and psychological challenges of facelessness. They put the patients and their needs on trial.

While this is, of course, standard bioethical practice and indeed the way the field is structured given scarce resources and competing needs, this particular instantiation of evaluating the patient allowed for a more concrete and ultimately even more damaging kind of patient adjudication: the first recipient of the face transplant, Isabelle Dinoire, was denied the right to privacy and very much put on trial.[15] Her doctors were too, as is to be expected for pioneering practitioners. Dinoire, however, was a victim of the narrative that questioned her need and, indeed, her right to have a surgery that gave her back her face. Some of the reasons why she and her doctors were subjected to such rigorous discussion had to do with the way this particular surgical event was communicated by the institutions involved. Subsequent surgeries were better framed and far less controversial.

The framing of the surgery by the French team has been carefully explored in a comparative context by Marjorie Kruvand.[16] Using an information-subsidies and agenda-building framework, she demonstrates that the media fiasco experienced by the French team was a result of poor public-relations planning. She contrasts this with the reaction three years later to the face transplant performed at the Cleveland Clinic. Kruvand argues that the PR team at the Cleveland Clinic dictated the terms of the discussion with a focus on medicine rather than morality, solidified by very careful media training and press-kit preparation for the physicians and others who would be talking with the press. I follow Kruvand's lead to look more closely at the results of the lack of media strategy on the part of the French team, which allowed morality and bioethics to emerge as the dominant discourse.

Other communications scholars have also considered the Dinoire case from various perspectives. Joy Cypher uses a disability-studies framework to explore the media coverage of the surgery as it dealt with the question of identity and body essentialization.[17] Cypher posits the ambiguity of classification of the procedure but focuses more on the implications for the normative body than on the shift in coverage and attitude toward the intervention.

Marc LaFrance considers the issue of identity from the perspective of Dinoire herself, exploring the impact of the surgery on Dinoire's own sense of self and the implications for transplantation and integration more broadly.[18] He raises psychoanalytic questions about the unity of the self and the insight that bodily substitution lends to what it means to be human. These questions are ones that lie at the heart of the uncertainty around, and opposition toward, the surgery as represented by the moral rather than medicinal framework of debate and analysis.

Samuel Taylor-Alexander also draws our attention to the subject position of the patient.[19] He reminds us that patients are, in this context, experimental subjects for which the bioethical and psychological considerations need to be reconsidered and rescripted in this new arena. Following LaFrance, he reminds us to consider the daily experiences and struggles of the patient. This approach is important to remember in light of the backlash against Dinoire; although she consented to the surgery, she was hardly in the same position as the surgeons who operated on her and managed her care. Taylor-Alexander traces the framing of the face transplant as an ethical intervention through questions of quality of life, arguing that the face transplant called the meaning of medicine into question.[20] I take up this theme, exploring how this expanded notion of medicine played out in the media coverage in the immediate aftermath of the surgery.

Recent scholarship has considered the evolving relationship between journalistic gatekeeping and the pressures of clickbait.[21] The ongoing debate about the role of agenda setting in the framing of news reflects the underlying assumptions about the consciousness of journalistic motivations, a question complicated by the financial considerations of the digital arena.[22] Herman and Chomsky's notion of flak provides a feedback loop through which journalistic framing narratives can change in response to audience concerns and interests; clickbait provides another such mechanism, and the two can work together as stories become refined over time. While these models are useful in considering the emphases placed on journalistic offerings, there is often more ambiguity than these approaches suggest. As a classic work by Gaye Tuchman has shown, framing is a repetitive process; journalists draw on older frames and incorporate emerging stories into these models as a way to deal with the breadth of information.[23] Journalistic practice incorporates the new information with more relevant or useful frames, sometimes causing the emphases on the stories to shift. Frank Durham highlights the failure of frames when there is a gap between the ideological commitment of journalists and their official sources.[24] He shows that sometimes reporters and investigators fail to agree, and journalists struggle to find an appropriate frame. The Dinoire case highlights the evolution of the frame; as a novel procedure, the face transplant had no entirely overlapping frame, causing journalists to turn to both cosmetic surgery and transplants and to think the procedure in terms of both, thereby tacitly supporting an expansive view of the medical enterprise as being physically and psychologically therapeutic. The bioethical and quality-of-life narratives provided the means to bridge these two procedures within the discourse around the face transplant. With the addition of information that destabilized the triumphalist bioethical narrative, journalists were able to more easily fit the story into an exclusively cosmetic framework that condemned the surgery from the very bioethical perspective that originally bolstered its acceptance.

A BREAKTHROUGH OF THE NONTECHNICAL KIND

The Dinoire story was big news, with widespread international coverage.[25] Drawing on newspaper reports in the United Kingdom, the United States, and Australia (three countries that were deeply involved in the face race), I observed a clear shift from initial cautious optimism about the possibilities of the procedure to an ostensibly bioethically based condemnation.[26] The second—and rather quick—narrative is rooted in the use of the cosmetic frame and in the broader approbation of elective facial manipulation as a form of social cheating.

The topic of facial transplants received a smattering of coverage in the press from 2004, when the Cleveland Clinic first received IRB approval for the procedure. As one team at the University of Louisville and another in London geared up for their own applications for approval, there was additional discussion of the benefits and ethical questions surrounding the surgery.[27] These were thoughtful discussions, and, as Taylor-Alexander and Kruvand have outlined, strategic ones, which followed the lead of the surgeons themselves to gain support for FAT in advance of its first iteration.[28] Such articles were to be the first of many in the press, in bioethical journals, and in the medical literature. The discussions were happening, and the issues were being debated, at a carefully managed pace, the nature of which I explore in the second half of this chapter.

And then Isabelle Dinoire attempted an overdose and had her face mauled by her dog. Within six months, she had a new face. (As a point of contrast, the first transplant recipient in the United States, Connie Culp, had been living with her disfigurement for four years and had undergone at least twenty-seven reconstructive surgeries prior to her transplant.)[29] Dinoire's timeline, particularly compared to the other institutions gearing up for the surgery, represented a quick turnaround. It was quick for the journalists too, and they hadn't quite worked out their reporting strategies yet. It was the perfect storm of controversial factors: an experimental procedure with unknown social and psychological implications, performed by a doctor with a headline-seeking history on a patient with a suicidal history. What could observers possibly find to debate?

A whole lot, of course.

The news first broke on November 30, 2005. The reports were vague, technical, and very careful, reporting the details of the surgical team, the cause of the recipient's disfigurement, and the novelty of the surgery. The *Evening Standard* noted that "[s]cientists around the world immediately hailed the operation as one of the greatest medical breakthroughs of the 21st century." They were careful also to note that "[t]he French ethics authorities . . . said that the procedure was too risky when doctors applied for permission last year but left the door open for the nose and mouth 'triangle' to be transplanted."[30] A brief technical overview of the procedure itself followed the details. Reports trickled in on the first of December as journalists started to get a handle on the story. The pattern that quickly emerged echoed the first article: a mention of the unprecedented nature of the surgery, followed by a brief discussion of the ethical issues and a technical explanation of the procedure. The *Guardian*, for example, opened with details about "the world's first face transplant." The article moved on to note that "[t]hese risks had to be balanced against the benefit the patient would receive" as part of a brief discussion of ethical objections. The article closed by answering the question "[i]s it a complex operation" with the claim '[n]ot especially, as transplants go."[31] The *New York Times* offered coverage of the story late on December 1 that ran along similar lines.[32] The article opened with the details of the location of the surgery, which had been done "for the first time." They moved to a discussion of risk, delving a little more deeply into the process by which the procedure was approved by the French ethical committee.

The first newspaper mention of cosmetic surgery in the context of the face transplant, on December 2, was a quick rejection of the comparison. The *Sydney Morning Herald* turned to an article in the *Journal of Medical Ethics*, itself written by surgeons advocating the procedure.[33] The *International Herald Tribune* noted that the article argued that "there was no need for a special ethical debate about face transplants, because they were not optional, cosmetic, surgery but a critical procedure for seriously deformed people who otherwise could not function in society."³⁴ The cosmetic frame was not deployed here, nor was it immediately adopted by other reports. It would, however, reemerge fairly quickly as the story progressed and reporters had new information to process and present.

As more details surfaced and time passed, press outlets found appropriate experts to interview and decided which aspects of the surgery to emphasize. The bioethical narrative, which reporters initially mentioned only briefly, soon developed into a major focus. The cosmetic surgery framing, which provided a broader context for the bioethical discussion, followed shortly thereafter. It was the combination of these two narratives—the bioethical and the cosmetic—that caused the tone of the discussion to coalesce into a condemnation of the surgery as not only dangerous but also somehow cheating.

The bioethical discussion had, as I've noted, been present from the outset. It quickly overcame the novelty of the surgery and its technicalities as the main focus of the articles. By December 2, the *International Herald Tribune* (and the *New York Times*) opened its article by writing that "[f]ace transplants are among the most disputed frontiers in transplantation science because they are so risky and because no one can predict what a patient who receives a transplant will look like after the surgery."³⁵ They were not alone in shifting the emphasis to bioethics at the outset of the article. The *St. Petersburg Times*, for example, opened their article noting that "[a]n ethics debate broke out over the world's first partial face transplant Thursday."³⁶ But the narrative itself was still unfocused, rehashing older bioethical discussions leading up to the surgery. The main thrust of the bioethical critiques centered on the surgeons and their desire to win the face race. In seeking an expert opinion, many articles turned to French surgeon Laurent Lantieri, an outspoken critic of Dubernard; Lantieri went on record early as saying that "Dubernard 'wanted to be first' to do a face transplant."³⁷ The strong implication was that Dubernard rushed in too fast, as the *Daily Telegraph* explicitly stated: "Some French doctors have accused the transplant team, overseen by Prof Jean-Michel Dubernard, of putting their desire to achieve a 'first' above the interests of the patient. Critics have suggested that normal facial-reconstructive surgery should have been attempted."³⁸ Dubernard was put on trial early in the process as reports and expert interviews critiqued his oversight methods and the speed with which Dinoire (whose name had not yet been released) received her operation.

Dubernard had anticipated these critiques, noting in his initial press conference that "'I know what is going to happen. We are going to be massacred by certain media.'"³⁹ To that end, as the *Telegraph* reported, "extra precautions were taken in an effort to avoid controversy."[40] Dubernard and others spoke in their own defense, noting that "as doctors, if we have the possibility to improve our patient, that's what we can do."[41] In a widely quoted statement, Dubernard said that "when he saw the extent of Mme Dinoire's disfigurement, 'I no longer hesitated for a second.'"[42]

By December 3, Dinoire's name and personal details had been released. The very first reports about her were sympathetic to her plight, lending implicit support to the surgery that helped alleviate the challenges she would face due to her disfigurement. On December 4, the *Sunday Times* (London) wrote that "Dinoire said that she had been living as a virtual outcast before being put on a waiting list in August for the operation. 'I had to cover myself with a mask all the time,' she said."[43] Many news outlets reported that her first words were "Thank you," or "Merci."[44] Dinoire's identity also provided reporters with details that allowed them to delve beyond the relatively dry technical information and somewhat vague implications of Dubernard's haste. Given the personal pathos of Dinoire's condition, reporters at first tread lightly around critiquing her, and even backed off the doctors by allowing them to speak to the dire nature of the situation. There was a strong sense of support for Dinoire's situation, echoed in a delicacy around her place in the bioethical debates.

But, as more information emerged, the tone shifted again. At issue was Dinoire's attempted suicide, which her doctors had initially denied and which both she and her daughter confirmed.[45] While some supporters in the surgical community remained committed to the procedure, as the *New York Times* wrote on December 6, "Ethical Concerns on Face Transplant Grow."[46] The details around rejection remained the same. What changed was the stakes for the mental-health implications, given Dinoire's suicidal history. With that new piece of information, the discussion shifted from the physiological risks to the psychological ones. The surgery itself changed in character; as a psychological intervention (or an intervention with serious psychological consequences), cosmetic procedures were now firmly in the frame. This made it appropriate to question Dinoire's own motives, and, in some cases, to condemn them. The *Daily Mail* asks, "Isabelle's new face: is it moral?"[47] But what it is really asking is if Dinoire herself is moral. The body of the article notes that even as Dinoire "was having difficulty eating, drinking, and speaking . . . she continued to smoke." The article concludes by noting that "[t]he 'face transplant' story raises the prospect that this could become the ultimate in cosmetic surgery. Imagine wealthy crones buying the face of beautiful young donors."[48]

And so? Technical objections aside (the cost, the scarring, the international attention on each transplant patient), is the problem that wealthy crones would no longer look like crones, or that they would look like the particular young people whose faces they bought?[49] Is this another objection to organ trafficking, made worse by the gratuitous nature of the transaction?[50] While the ostensible issues revolve around the figure of the donor and her exploitation (which would be serious indeed, though even within the organ sales debates there are voices on both sides), yet again we see that faces are a special case.[51] In particular (because people manipulate themselves to look younger all the time), we see that taking on the faces of others are a special—and specially vexed—case. People with facial disfigurements are held to a different standard, not only because the condition itself will not directly kill them but also because we simply can't make sense of what a person with the face of another is. More specifically, we can't make sense of *who* someone with the face of another is. But here's what we seem to know: she should at least deserve it.

Dr. Eileen Bradbury, a British consultant psychologist specializing in plastic surgery patients, noted in the *Independent* that "[y]ou would have to be pretty desperate to have this sort of surgery, which carries so many risks. Should someone that desperate be submitted to it?" Underlying the general suspicion of seemingly cosmetic interventions is the specific suspicion that Dinoire, whose need stemmed from an event that was surely her fault and was at best careless and at worst insane and immoral, did not deserve it.[52] She did not deserve to get a new face. Like a woman discussing her own cancer-related facial disfigurement with the *Sunday Times* (London), Dinoire should have suffered with her lot: "I've thought a lot about facial transplants, and it sends shivers down my spine."[53] For this woman, Susan, the fact that her surgical interventions all originate from her own body is of the utmost importance. She does not disallow improvement attempts, but she draws the line at those that she sees as compromising her fundamental selfhood: "Everything that I am today is part of me, grafted from my back or my leg. But I'm still the same person. Face transplants are not you, so you don't know psychologically how you'll react." She takes pride in having made peace with her situation, arguing that others should as well: "I'm known for the way I look. Rather than trying to eradicate our disfigurement, we need to accept it and get on with life."

It was at this point that the headlines began to shift. In addition to the editorials, which were explicitly critical of the ethics of the operation, reports began to focus on Dinoire's mental health and the motivations of her doctor. The *Guardian* entitled its December 5 article "New Row Breaks out over

Face Transplant: Ethics Professor Attacks 'Lack of Consultation'; Donor and Recipient Had Both Attempted Suicide."[54] A day later, the *New York Times* reported on "A Pioneering Transplant, and Now an Ethical Storm."[55] Additional information emerged about Dinoire's financial stakes in the operation; as the *Times* (London) reported on December 8, Dinoire "signed a deal that could make her tens of thousands of pounds from the sale of photographs and a film of the operation."[56] A victim, in the eyes of the press, no more.[57]

The first version of the media story about Dinoire was a cautious one, recognizing the advantages of the surgery but fearful about its applications and implications. The surgical story held popular interest, briefly. But Dinoire's personal history was far more controversial and generative, as were the characters and motivations of her doctors. As the focus of the media coverage shifted from the medicine to the ethics, the tone shifted from discussion to debate, and from debate to public-relations debacle. Dinoire and her doctors were on trial, and, while they may have won the face race, they lost face.[58]

This is the second half of the story. But how did we get here? We've seen the broad history of transplants and their relationship to bioethics in chapter 2, and we've seen how the face transplant is both like and not like whole-organ transplants and cosmetic surgery. But there is a better analogue, not just theoretically but medically. Hand transplants—another, and prior, visible-organ surgery that was not necessarily lifesaving—in many ways foreshadowed the conversations around face transplants. And in many ways, despite the medical, technical, and even cultural similarities, they did not. Because despite the commonalities, faces are different, and we need a different language to talk about them than traditional transplant frameworks. (We're still figuring out what that language might be; psychology talk was too diffuse and meant different things to different people.) As medical anthropologist Linda Hogle has been quoted, "The face is the most intimate, most individual characteristic of your body. It's who you are."[59] Faces are special.

BRAVE NEW WORLD

The first hand transplant was attempted in 1964 in Ecuador and was rejected after two weeks.[60] The first successful hand transplant was performed on New Zealander Clint Hallam in Lyons in 1998, by Jean-Michel Dubernard, who ultimately was the lead surgeon on the Dinoire case. Well, it was the first *technically* successful hand transplant; Hallam said that "as soon as I saw it I didn't like it, as "the donor hand was bigger than my hand, bald and pink, and my skin is olive-toned and with hair. It didn't match."[61] Claiming a negative reaction to the drugs, Hallam quickly became immunosuppressant noncompliant, causing his hand to wither and "hang uselessly by my side." Given that "it looks hideous because it is withered," he did not "see any point in keeping it any longer."[62] Hallam begged for its removal against the advice of his initial surgical team, who called the operation a total success. He finally convinced British surgeon Earl Owen to remove the rejected hand, two and half years after the original operation.[63]

Hallam's reaction to his procedure set the stage for later discussions about psychological rejection and precipitate experimental surgery. The procedure itself set the stage for the face transplant; following the first successful hand transplant in the United States a year later, a member of the transplant team suggested that, based on the outcome of this surgery, the technically and immunologically similar face transplants would follow in under a year.[64]

As we've seen, it took a little longer than that (about seven years total) but not because of the technicalities of the surgery or even the medical risks it might pose. It was the ethics of the procedure that stalled its implementation; IRB (Institutional Review Board, the body attached to hospitals and academic centers in the United States that approves human-subjects research and procedures) and ethics-board approval (internationally) was a long time coming, with strong voices arguing that, from a risk-benefit analysis, the surgery was simply not worth it. We're going to explore those voices through the medical and bioethical literature and try to decode what was really bothering them.

For some, the very debate—heated and, for a long time, tilted in favor of the naysayers—was itself confusing: the opportunity to give the faceless back their lives seemed like an obvious and worthy one to pursue. It should be done thoughtfully and carefully and with informed consent and psychiatric supervision but without bioethical angst. People would be given the renewed ability to talk, taste, smell, experience touch, and go out in public without causing mass hysteria or at least being the subject of stares, screams, or tears.[65] (People deserve to be able to go in public without others staring, screaming, and crying at them, especially if they didn't opt into the condition that inspires these reactions.) Others—from the moment the surgery became a theoretical possibility following successful hand transplants, to the first IRB approval granted to the Cleveland Clinic in 2004, to the first partial surgery performed in 2005 at the Centre Hospitalier Universitaire Nord in France—disagreed. These claims will be familiar by now. Many questioned the need for these supposedly nonlifesaving surgeries, given the complicated immunosuppressant regimes (and the associated disease risks) required by the recipients.

These critics challenged the possibility of true informed consent by patients whose lives were so compromised by their appearance. (In their critiques, they did not prioritize the quality-of-life stakes balanced between these problems.) They wondered about the psychological implications of a medical procedure that would give recipients the face of another (as opposed to no face at all). They interrogated the motives of doctors eager to engage—too soon—in pathbreaking surgeries without properly considering the risks. They worried (drawing on cinematic and literary tropes of the evil/mad doctor) if medicine was going too far.

The ones posting these objections were bioethicists, mostly, and they mostly thought the surgery in terms of the cosmetic framework. And that, mostly, made it not worth the risk. On the other side were the surgeons, who slotted the procedure under the heading of therapeutic, and thus, necessary. It's a neat and basically clear binary. But I'm less interested in binaries and more interested in the space that lies between. Which leads me to ask of both the bioethicists and the surgeons debating the surgery what the stakes are for their particular positions, and what it might look like to allow the surgery on different terms, terms that explode the binaries between cosmetic and lifesaving, between virtuous and gratuitous, and, maybe, between you and me, yours and mine.

As I analyze the technical discussions of doctors and bioethicists, I argue that the deliberate (and later widely copied) framing of the surgery as cosmetic obscured the real source of the problems that bioethicists had with the face transplant: a deep suspicion of rendering the indexical relationship between face and character unstable and vexing the 1:1 correlation between face and specific individual.[66] Or, if we can't know others through their faces, how can we know them at all? Which begs the question: who is facing the ethical dilemma—surgeons and patients or the society that has to relearn the meaning of the face?

THE DEBATES BEFORE DINOIRE:
TECHNICAL AND BIOETHICAL

Couched in broadly bioethical terms, the debates around Dinoire's surgery were located largely in the popular media. They focused less on the medicine than on the characters involved, putting Dinoire and her doctors on trial. While the newspaper reports echoed many of the concerns in the bioethical literature, they paid more attention to the specific surgery than the procedure in general, speaking both to the novelty of this event and the poor framing and PR effort by those involved in its release. Dinoire's doctors were roundly criticized for attempting to win the face race by using an unsuitable candidate who was improperly screened. Those were the specifics. There were also the general issues which concerned not the technical aspects of the surgery but its long-term psychological impact on the patient and, equally (though more subtly), the long-term impact on society at large. What monsters were being unleashed by this apparently inevitable but suspect surgical innovation? It was really happening, finally. The fears weren't new; the medicine really wasn't new. And the doctors and the bioethicists knew that all along.

According to medical and newspaper sources, the surgeries "could have been done anywhere where there are trained microsurgeons and plastic surgeons—China, Australia or other countries—from a technical point of view." Medical developments had not been the limiting factor to date; rather, "[w]hat has been holding it back are ethical issues."[67] These ethical issues were discussed in great detail in the professional literature well prior to the first actual surgery. There is a clear split between the largely supportive surgeons and the mostly skeptical bioethicists in their approach to the procedure, which mirrors the categories in which people tried to place the surgery. For most bioethicists, the face transplant was (more like) cosmetic surgery; for surgeons, it was medical, therapeutic, even lifesaving, and akin to whole-organ transplants. This is predictable; the professional success and indeed personal research interests of surgeons are furthered by new procedures and their results. Surgeons also work in direct contact with the patient population and bear witness to the challenges they face. They, better than anyone, know the medicine and have a sense of the technical risks. But, like everyone else, they could not predict the psychological ramifications for patients and society at large. And like everyone else, they tried to slot this category-resistant procedure into neat boxes, failing to imagine the new possibilities of the bodily imaginary the surgery could support. Now *that* would have reframed the debates entirely.

The professional discussions around the ethics of facial-allograft surgery were framed at the outset by the recognition that, despite parallels with other kinds of transplants (particularly hand transplants, the most recent and technically similar procedure), this intervention would be unique and present a unique set of psychological considerations. To explore these issues, I turn back to 2004, when the cadaver and animal studies of the surgery had progressed to the point of yielding meaningful results. It was these results that led Maria Siemionow's team at the Cleveland Clinic to begin preparing IRB applications for permission for the surgery. And the ethicists responded.

While this is not a comprehensive survey of all of the bioethically related articles about the procedure, I have highlighted those that best represent the various issues at stake at that time. Some articles, usually situated in specialty surgical journals and written by surgeons invested in attaining permission for the surgery, outline possible screening procedures and review the research to date. Others, published in more explicitly bioethical forums, raise concerns about the procedure and, following its first completion in 2005, critique the process to date and offer suggestions going forward. Of course, the writers of these pieces enter the discussion with their own concerns and stakes in the outcome, particularly the surgeons who were preparing their IRB and ethics applications even as they were arguing for its approval in the professional debates.

In their 2002 *Lancet* article "Face Transplantation—Fantasy or the Future?" Shehan Hettiaratchy and Peter E. M. Butler offer one of the earliest medical defenses for proceeding with the possible surgery.[68] Hettiaratchy is a British plastics and reconstructive surgeon, and Butler was at the forefront of face-transplant research in the United States and in the United Kingdom, where he is director of surgery and trauma at the Royal Free Hospital in London. The UK team seemed on the cusp of joining the early pioneers in the surgery, receiving ethical approval in 2006, but a series of setbacks keeps them frustrated in the process as of this writing. Butler has met considerable resistance from face-deformity advocates in the United Kingdom, including Prof. Iain Hutchison, founder of Britain's Facial Surgery Research Foundation. Hutchison argues that: "There is something quite fundamental here that creates a difficult ethical conflict. When you carry out a facial transplant, you are converting a healthy person with a disfigured face into an unhealthy person who looks more normal."[69] "Healthy" here refers strictly to physiological health in this assessment; mental health has no place in Hutchison's evaluation. Nor, as we'll see, did it in the evaluations of many others. Hutchison acknowledges only a narrow definition of health and treatment. And these parameters matter: if mental health is not at stake, then the transplant itself cannot, according to most bioethical risk-benefit evaluations, be allowed. Butler's response to Hutchinson and others has always been to turn attention to the needs and desires of the patients, who "live a terrible twilight life, mostly shut away and hiding from public gaze" and deserve to have the chance to leave their homes and interact with others.[70] Butler has been a consistent voice supporting the surgery from its earliest days of feasibility; he has yet to lead a face-transplant surgery himself.

In this early piece in one of the foremost medical journals, Hettiaratchy and Butler argue that both the immunosuppressant concerns and the technical issues of face transplantation are no longer barriers to the successful completion of the surgery. While they acknowledge that "the concept may be shocking" and "appears to have come straight out of science fiction" (note the turn to fiction—one of many—as a reference point for how people approach the procedure), the benefits outweigh the risks for those who "have serious physical and psychological problems that cannot be solved by conventional treatments." For these early advocates of the surgery, the psychological stands equal to the physical as compelling reasons to proceed with the intervention.[71] Not only is the surgery permissible, they argue, but "[i]f face transplantation is shown to be the only effective way of treating these severely disfigured patients, then doctors would have a duty to use the technique."[72] This is an unequivocally strong stance that does not consider the surgery experimental, either physically or psychologically, and indeed draws no real distinction between the two, seeing psychological dangers as just as unthreatening as the physical ones, particularly in the case where doctors have the ability to alleviate these challenges through surgical intervention. This is a particular stance and far from a universal one: many doctors would balk at performing possibly risky surgeries to satisfy the nonphysical demands of the patient. (Others wouldn't. And haven't.)[73] The authors here consider the risks of not doing the surgery both transparent and extreme; that is itself up for debate. Assessment of psychological need always is.

A 2006 article by Siemionow et al. in *Plastic and Reconstructive Surgery* is another technically based case for the surgery.[74] Maria Siemionow is director of plastic-surgery research and head of microsurgery training at the Cleveland Clinic and received the first IRB approval in the world for the procedure. She was not, however, the first to perform a face transplant: Siemionow took her time to lay the ethical, political, and social groundwork for the procedure in a series of technical and bioethical articles that she published and placed carefully in the years leading up to and following her IRB application.[75] She was successful in that her first surgery, on Connie Culp in 2008, was far less controversial than the Dinoire procedure. She was unsuccessful in that she was, in the end, not the first. She lost the face race.

Siemionow et al.'s 2006 article, based on a series of anatomical dissections on cadavers, outlines the superiority of facial allografts over facial-flap approaches, highlighting the situations in which skin grafts and other approaches simply could not achieve the same results as a facial transplant. The authors note that "perfect match of facial skin texture, pliability, and color can only be achieved by transplantation of the facial skin allograft from the human donor."[76] All other avenues, the article claims, had been exhausted; for patients in particular situations, only the allograft itself could achieve substantive results.

The article is careful to note that, although it "would be difficult to predict the appearance of the face following transplantation," it was unlikely that recipients would look more like their donors than themselves. This conclusion was based on computer modeling, which "suggested that the face would carry more of the characteristics of the recipient skeleton than of the donor soft tissues."[77] (They were, as subsequent actual transplants show, entirely correct.) While the question of appearance was not strictly germane to the study in the article, the authors strategically tried to alleviate fears of creating visual clones or opening the door to elective facial allografts. (They didn't really succeed. Nor did they address the underlying issues of vexing the unique correlation between one face and one person.) They placed this discussion directly prior to their schematic list of the remaining (nontrivial) obstacles to facial-tissue transplantation, which included achieving IRB approval, finding appropriate recipients and donors, technical feasibility, immunosuppressant protocols, and, last in their list, "responding to the ethical, social, and psychological issues."[78] This last piece would turn out to be an ongoing project to this day, even following the achievement of all the others. The first requirement, IRB approval, was granted to Siemionow significantly before the other pieces were resolved.

As part of her campaign for IRB approval, Maria Siemionow accompanied her technical article in favor of allograft surgery with a more explicitly bioethical piece, which she cowrote with George J. Agich, a philosopher of clinical ethics. This paper was published in the *Journal of Medical Ethics* in 2005 and provocatively entitled "Until They Have Faces: The Ethics of Facial Allograft Transplantation."[79] This strongly argued article leaves no doubt about its stance, opening up with a position statement immediately below the title that reads: "Let us not forget those whose quality of life will be enhanced by this technique."[80] Agich and Siemionow's frustration with those opposing the process is clear even in the abstract, which notes that "[t]he ethical discussion of facial allograft transplantation . . . has been one sided and sensationalistic." The opposition to the procedure, they claim, emerged from "film and fiction rather than science and experience" (film again!) and, in so doing, "overlooks the plight of individuals with severe facial deformities."[81] Throughout the course of their analysis, the authors claim that the media reliance on film to draw conclusions about the surgery frames the procedure as "cosmetic or nefarious," which in turn mischaracterizes the patients' real "clinical need."

To that end, "the suffering of individuals with severe facial deformities is trivialized as negative social attitudes marginalize the right of these individuals to choose investigational procedures to improve their situation."[82] Overall, they argue that detractors of the procedure hide their own discomfort and qualms, fueled by media sensationalization, behind the rhetoric that the procedure would be too psychologically disturbing to recipients and their loved ones. According to this piece, this approach fundamentally misunderstands the depth of trauma involved in facelessness and the vast clinical and quality-of-life improvements that would be enabled by facial-allograft transplantation.

The authors systematically discuss and discount a wide range of objections to the surgery. They start with the ostensibly easily dismissed concerns, including the possibility of psychic or personality transfer, which they deal with by noting that "[t]ransplantation of the skin envelope would not transfer the donor visage . . . transfer of any significant sense of facial identity is not a realistic goal."[83] As we've seen in chapter 2, identity transfer was a serious concern for some. Indeed, Françoise Baylis makes a case for identity transfer posing an ethical challenge to the surgery in a 2004 article in the *American Journal of Bioethics*.[84] The more complex issues are those that are hedged on uncertainty, including adherence to immunosuppression regimes and the ever-present question of the true need for the surgery. To both claims the authors make a strong case, noting that the reality of "a life of isolation" outweighs concerns about adherence to medical regimes, particularly when the quality of life is so dramatically enhanced.[85] The quality-of-life calculus is always considered in the not strictly lifesaving cases of cleft palate and kidney donation, both of which carry risks of rejection and may require lifelong medication regimes. To emphasize the quality-of-life argument beyond questions of functionality and the sheer aesthetics of the issue, they point out that the very ability to communicate is fundamentally compromised without a face: "face to face communication remains the paradigm of communication. Because of its expressive function, the face carries with it important symbolic, social, and psychological significance that cannot be overlooked."[86] Except, through media sensationalization, overlooked it often is. Underlying this willingness to overlook the situation of faceless patients is observers' own discomfort with interventions, which, the authors claim, "reflect[s] the conviction that people suffering from severe deformity should simply endure their condition; their suffering is apparently not compelling." They conclude that "[t]he sad reality is that until they have faces, society does not seem ready to countenance these patients."[87]

Both Siemionow et al. 2004 and the article published shortly after by Petit et al. in the same journal referenced the 1998 hand transplant as the groundbreaking immunological precedent, as it was "an immunological challenge but not a surgical challenge."[88] Doctors early acknowledged that the technical issues around both hand and face transplants had been readily conquered, and, with the innovations in hand transplants, the relevant immunological issues had also been worked out. What remained, as these articles discuss, were significant institutional, ethical, and sensational issues, all of which could be addressed from a variety of perspectives. The Petit et al. intervention situated science as the relevant framework by which to evaluate the status of and need for facial-allograft surgery, engaging in the article's introduction in both professional stake claiming and careful rhetorical framing: "In the present article, we seek to review the current status of science regarding the prospect of face transplantation and then detail the reasons why the first human face transplantation could or could not be performed."[89]

This piece carefully distinguishes between the debate within the medical community and that within the media, which offers "little help" in answering the relevant questions. The article contends that the "prospect of face transplantation has been sensationalized by mass media" (little did they know), turning the potential procedure and its possible recipients into a "kind of 'scientific reality show'" (and another one . . .) that "diminishes the potential value of these procedures." The medical community, by contrast, was concerned with "scientific interest," wondering if the science had advanced to the appropriate state to proceed (an obvious softball given that the first half of the article outlined precisely why this was no longer a concern). Science was of course not the only issue; the two more unanswerable questions included whether this procedure would benefit the patient and the more self-serving concern that "the procedure [would] affect people's opinion of doctors in general and the practice of transplantation in particular."[90] While posed as pressing questions, all of these issues were dispatched in the course of the authors' survey of the state of the field to date, leading to the inevitable conclusion that "[f]ace transplantation should be pursued as a potential solution for a small and selected group of patients with conditions that cannot be adequately addressed by conventional reconstructive surgery procedures."[91] Note the insistence on mapping out only a very contained group of potential recipients; this was not, the authors underscored, elective medicine for the masses.

Not everyone concurred with these (somewhat self-serving?) surgeons. And it wasn't only the "sensationalizing media" that disagreed. And disagreement did not rest only on the basis of the science, the reputation of the doctors, or even the presumed benefits to the patients. Enter the bioethicists.

Of course, (as we have seen) the bioethicists had been part of the transplant debate for a very long time, even before they were bioethicists as such, given the relationship between organ transplants and the birth of the field.[92] They certainly weighed in prior to and following the first (ultimately rejected) successful hand-transplant surgery, and the many double and single transplants that followed. Despite the overwhelming number of compliant patients, that pioneering recipient in 1998, noncompliant Clint Hallam, disproportionately influenced the discussion about face transplants and their risks and benefits. Once again, the necessity of the surgery was weighed against the dangers, with little consideration of quality of life or therapeutic benefit as a factor in the calculation. The issue of quality of life and the very real therapeutic benefits of transplantation were often ignored or rendered marginal, making the risk-benefit analysis significantly lopsided against the surgery. As Tia Powell notes in her 2006 review article "Face Transplant: Real and Imagined Ethical Challenges," "[i]n several cases, psychological factors appear to be the main reason for the negative recommendation, yet the experience and preferences of potential transplant candidates seem strangely ignored."[93] Ignored, I would add, by the majority of bioethicists. The surgeons, who were in many cases in direct contact with the potential recipients, drew more directly on the stakes of facelessness.

Part of the problem was the very real difficulty of evaluating the nature of psychological need. (This is not a new problem.) In the case of the risk-benefit analysis of the face transplant, the risks are such that the need has to balance out the dangers. And in addition to the need for a new face (psychological stress) was the psychological stress that would come with a new face. The (projected) stress was the part of the reason the Royal College of Surgeons called for further research into the procedure in 2003.[94] They dwell a great deal on psychological issues, noting that "the extent of psychological distress resulting from a visible difference is not well predicted by the extent or severity of the disfiguration." There is a variety of responses, such that "[s]ome cope well with an extensive and very visible disfigurement while others struggle to deal with a relatively minor difference." They note the implications for potential recipients, given that "[p]aradoxically, the more vulnerable will be less equipped to deal with the aftermath of complex transplant surgery, uncertain outcomes, and ongoing treatment."[95]

As various institutions began to explore the issue and prepare IRB applications, the *American Journal of Bioethics* devoted an entire special issue to the question of facial transplants in 2004. The articles on the whole are more supportive of the surgery than those in other bioethical publications, if cautiously so. This support makes a great deal of sense, given that the special issue coalesced through the initiative of the cutting-edge surgical-transplant team at the University of Louisville, which had performed the first US hand transplant on Matthew Scott in 1999. As the introduction to the issue explains, "[a]s the team there began to consider attempting the first face transplant, they wanted to identify and consider any ethical issues that they had not anticipated. The result was a process by which they explored what they identified as the major ethical issues . . . they have decided to expose their thought process to the wider bioethics community." "The result," chronicled in the special issue, "is a lively and topical debate about how and whether to proceed with this procedure."[96]

Consistent across all these discussions is the challenge of informed consent, which is one of the key issues that both the Royal College of Surgeons and the *American Journal of Bioethics* special issue address and with which both surgeons and bioethicists struggle. At stake is the particular challenge of informed consent in cases of extreme suffering and desperation, when potential patients may be willing to do just about anything and cannot properly consider the risks. A 2005 article by Rhonda Gay Hartman in the *American Journal of Law and Medicine* strives to consider the entire range of issues around face transplants, including the following questions: "how should this innovative surgery be conceptualized and what would it entail; for whom would it be available (and deemed appropriate); what would be the physical and psychological risks to recipients; how might it impact recipients and families, as well as donors and families; what should informed decision making about receiving and donating involve; how should society influence and respond to strides in biomedical technology that transcend a core of human existence; and whether legal policy and regulatory oversight are desirable."[97]

Hartman starts by asking about "the value of the human face," which she notes is "a vexing question with no simple answer."[98] Simple answers are, rightly, in short supply for Hartman, who repeatedly emphasizes the unknown and unknowability of many factors in these as yet theoretical surgeries, including patient compliance with drug regimes, questions of risk, and the uncertainty about what legal regulations would govern the surgery and the donors and their families. Also unknown, Hartman posits, is the possibility of informed or voluntary patient consent in cases of such extreme desperation. To Hartman, "any concept of informed decision-making in this context seems counterintuitive; a facially disfigured person's euphoric anticipation, accompanied by rekindled hope and expectation, militate against voluntary decision making, an element crucial to the legal and bioethical contours of informed medical decision making."[99] Hartman elaborates that "personal choice should result from one's own values unencumbered by coercive influences," an impossibility in this case because "the face closely correlates with identity in a constitutive sense," which "suggests an inherently suspect decision." The problem of consent is particularly vexing in any experimental intervention; "the 'nothing-to-lose' attitude of persons willing to undergo unproven, radical procedures in the hope of recovery arguably weakens one's capacity to decide whether to undergo risks."[100]

Hartman does acknowledge that potential face-transplant recipients can be thought of as similar to other "pioneering patients in organ transplantation," for whom "the values of autonomy and well-being [which are] underlying informed consent are not necessarily opposed." Equally, "contributing to the collective welfare" is an important consideration.[101] Hartman does not take a definitive position on this question, choosing instead to compare potential recipients to "those afflicted with incurable disease who choose to participate (and are permitted to do so) in pioneering research."[102]

This is a neat—and well rehearsed—rhetorical sidestep that renders the patient incapable of offering informed consent precisely because of the extremity of the situation, while at the same time allowing the patient to participate by virtue of that very extremity. As long as the patient is prepared to die, or is essentially the living dead, with no real expectation of the surgery beyond helping further research, then the consent being given is not tainted by the intensity of the patient's hope for recovery. While this is a kind of ethical solution to the problem, the very framing of the issue is itself predicated on a specific set of assumptions about the risks of the surgery that doctors themselves discount. This surgery actually isn't the same as pioneering cases of organ transplantation. The technical aspects of the surgery were not—even for Hartman—among the unknowns. The psychological aspects certainly were; this piece conflates the psychological unknowns with the technical ones, masking discomfort with the former under a discussion of the latter. While the psychological risks are not isolated in this piece, it is they to which the author refers; they are the true unknowns, and this is the true risk. They are also, unlike traditional physiological responses to transplants, not reproducible in quite the same way, rendering the research approach to this surgery a bit more complicated. Nevertheless, Hartman's article stands out among those written by nonsurgeons in its particular recognition that patients without faces suffer very real and in some cases life-threatening or life-undermining challenges beyond the traditional physical risks faced by whole-organ transplant candidates. In conflating the potential death of the original organ-transplant volunteers with the outcomes for potential face-transplant candidates, Hartman acknowledges the depth of the psychological potential consequences alongside the physiological ones.

Some surgeons changed their minds. A 2006 article in *Plastic and Reconstructive Surgery* describes the waning commitment of one doctor to the potential of facial-allograft surgery. Luis Bermudez outlines his optimism about canine research into the procedure following his frustration with the limitations of skin reconstruction for burn victims. He has since altered his stance and now "strongly believe[s] that this surgery should not be performed in a human being."[103] Like many other advocates, he bases his claims on hand transplants, which he finds to be far from ideal as a model, with significant clinical and functional drawbacks. He introduces an important and hitherto under-discussed difference between hand and face transplants; a failed hand transplant may leave the patient no worse off than prior to the surgery, as "the amount of the patient's tissue removed . . . to perform the transplant is not significant."[104] By contrast, "[i]n facial skin transplantation, the complete surface of the face would have to be removed to insert the facial transplant. If the transplant is rejected and lost, the final deformity would be worse than the one before the transplant."[105] The possibility of emerging from surgery with even greater loss is a significant risk indeed, and helps explain Hartman's contentions that participation in the surgery ought to be considered selfless engagement in future research rather than personally beneficial. The comparison breaks down, however; in some cases, hand-transplant candidates have to have parts of their arm—up to and including the elbow—removed prior to the surgery.[106]

In his objections and ethical considerations, Bermudez operates in the realm of scientific studies, accumulated evidence, and experience. He is largely uninterested in questions of the recipients' psychological stability (no psychology talk for him!) and the possible trauma of seeing the face of another on oneself, recognizing that such objections fall outside his explanatory framework and theoretical approach. Bermudez' approach elides the quality-of-life questions that may have undergirded his original support for the surgery, but he also does not draw on psychological risk as a reason not to proceed. Psychology talk cuts both ways; he engages in neither of them.

Bermudez echoes concerns about chronic rejection and the risks of immunosuppressant regimes, which compound the problems of unknowability presented by to date inadequate animal research on facial-transplantation procedures. These concerns are consonant with those of others and remain well grounded in the medical literature. Bermudez, like other detractors of the surgery, does not engage in depth with questions of quality of life, arguing that "the concomitant risks, such as rejection and immunosuppression," outweigh the benefits, "considerations [that] are especially valid in a younger patient with a normal life expectancy."[107] The emphasis on life expectancy underscores the lack of attention to life quality; longer life does not mean better life, and in extreme cases just the opposite might be true.

At least, that's what some say. A 2006 study explored how many life years three groups of people would be willing to give up (alongside other costs) in exchange for various transplant procedures. Across the categories, which included healthy individuals, whole-organ transplant recipients, and those with facial disfigurements, people were by far willing to accept the greatest risk (and loss of years) for faces; the value of life without a face, it seems, is significantly lower.[108] But we already knew that. Though for a while we seemed to have forgotten.

THE DEBATES AFTER DINOIRE, OR, A CASE STUDY IN BAD PR

In hindsight, the depth and vehemence of the condemnation of Dinoire's surgeons and the procedure in general is puzzling. We know now that greedy and manipulative doctors did not fundamentally alter human relationality. (Or did they . . . ?) We see the benefits and are less wary of the risks. We understand that psychological issues associated with seeing the face of another are dwarfed by the issues of seeing no face at all, especially compounded by the inability to eat, smell, feel facial touch, and, importantly, go out in public without being stared at in fear and horror. (This is an argument. Mine, but not just my own.) And surely the ability to adhere to an immunosuppressant regime has been well tested and explored in the long history of organ transplantation generally.[109] It is not merely the combination of these two kinds of risks that led to the discomfort with and condemnation of this surgery. Rather, it is the transplant of the face that motivated the debates.

Let me be clear: the body and its manipulation has also been subject to debates, many of which I outlined in the previous chapters. Organ transplantation has likewise been a source of lively discussion. Hand transplants generated a faint echo of the kinds of discussions we see with respect to facial allografts.[110] Very faint. The discussion about hands was not as extreme. This level of debate was something new entirely. And it was partly because of the sensationalistic nature of the particular story; it was also partly media mishandling and poor framing, partly timing, and partly personalities and events. But I still claim that the face has special status beyond all these specific factors. Changing the face is complicated. Swapping it is more complicated still: putting the face of one person onto another is a special bioethical, social, and interpersonal kind of problem, and one that people didn't know how to talk about, so they made it into something else, something more familiar, and something solvable. What that something was depended on the perspective of the person engaging in the debate.

Press coverage of the surgery ranged from technical explanations of the surgery to biographies of the doctors to speculation about Dinoire and her mental health and stability. By far the most common thread of discussion centered on the ethics of the surgery itself, whether it ought to have been done on Dinoire, and whether it ought to have been done at all. More specifically: were Dinoire's doctors irresponsible limelight seekers performing an ethically tricky surgery for their own advancement? And, at the same time, was the patient undergoing a (less) medically and (more) psychologically risky procedure whose costs outweigh the benefits? Underlying all of these debates was a rising concern bordering on hysteria that medicine had entered the realm of science fiction, of mad-scientist doctors and apocalyptic possibilities, of horror films (explored in the previous chapter) come to life. With this lens, the question of risk changes tenor: were the potential recipients or the larger society that has to grapple with their new faces and what they might mean the ones who were being made vulnerable?

While it is true that Dinoire could live without the transplant, it is dramatically untrue that in this (and later) cases, there was no therapeutic benefit. For example, with new lips, Dinoire was once again able to eat. Others who came after her regained the ability to smell, feel touch, speak, taste—some for the first time in decades.[111] The technicalities of the surgery had been explored and practiced in detail in prior facial-reattachment surgeries and hand transplants, as well as in simulations, animal surgeries, and cadaver experiments. These experiences helped surgeons develop expertise in the delicate connections needed to attach the face to sources of oxygen while working with the most minute threads and tissues. The other major set of medical concerns, around organ rejection, had been well worked out in the decades of accumulated experience of immunosuppression drugs for other transplants.

While this particular intervention had never been done before, the technical aspects were not themselves new. It was the narrative precedents that were groundbreaking, seemingly situated as they were in science fiction rather than medicine. And yet, as we have seen, the first organ transplants generated similar reactions, discomfort (and fear), and bioethical quandaries, in many ways contributing to the birth of that field.[112] But the field got better. Not doing the surgery, for most whole-organ transplant candidates (and even those with end-stage renal disease), meant certain death. Transplants—when possible, when organs, drugs, and operations are available—mean life. In the case of the face transplant, the trade-offs remain unclear, because of both an inability to predict what the new face might bring (and what kind of life it offers) and great uncertainty about the status of the surgery itself.

Here we have a kind of added pressure on survivors of facial trauma to cope with their situation, to be ennobled by suffering. While there is a strong trend to place this particular burden on people with disabilities, there is an even greater emphasis on embracing facial disfigurement rather than trying to change it, a legacy of the suspicion of cosmetic interventions more broadly. Facial disfigurement is not quite considered a disability, and it's a tricky kind of thing. Is ugliness a disability? The kind of disability that should, bioethically, be traded for long-term physiological risks associated with surgery and subsequent medical regimes? But even that question makes a set of assumptions about what facial impairment is and what its amelioration constitutes. But faces still pose a particular set of complications: is damage to the face a disability at all? Much of the response to Dinoire seems to indicate that hers was not or, at least, not enough of a disability that its correction was therapeutically warranted. Because while it was always clear that potential transplant candidates had no faces, it was not clear the extent to which they needed new ones. At least according to everyone but the first recipient: she had no such lack of clarity. She knew what she needed. As was widely reported, Isabelle Dinoire's first word after her surgery, when she first looked in a mirror, was "merci," or thank you.

Perfect clickbait. So was the rest of Dinoire's story, in another way entirely.

THE AFTERMATH

It all happened very quickly. Too quickly, perhaps, given the concerns about patient confidentiality and Dinoire's right to privacy. As Lengelé et al. expressed in their 2007 editorial polemic decrying the irresponsibility or, in their words, lack of "elegance" in the media portrayals of the first facial allograft, donor and recipient confidentiality is a central tenet of medical ethics and practice.[113] Dinoire's name was swiftly leaked, first by the British papers as early as December 3, 2005 in the *Daily Telegraph*, with an "exclusive" follow-up with pictures the following day, and then, with the most intimate details of her life and the circumstances of her disfigurement, by media around the world.[114] The French team who managed her surgery provided an unprecedented amount of information about her personal life, the details of her surgery, the visual spectacles of her face before and after, the operation, and even her hospital room.[115] By inviting the media so enthusiastically and with so little sophistication and planning, the French team set the stage for what Cleveland Clinic head of corporate communication Eileen Shiel later decried as a media fiasco.[116]

But, as Lengelé et al. point out, the media was unlikely to keep secrets. More disturbing to them, however, was the fact that the medical establishment proved equally poor at respecting Dinoire's and her donor's confidentiality. Given the controversy around the procedure, there was always going to be debate about whether the doctors should have proceeded. The unique factors affecting visible organ transplant meant that questions of psychological trauma were always going to be at stake. The lightning strike that sparked the inferno, however, was unique to Dinoire's own life story, dramatically raising the issue of personal fitness and stability. At issue was Dinoire's overdose, which caused her to remain unconscious while her dog devoured her lower face, possibly in a desperate attempt to wake her. (At issue were the circumstances by which her donor face became available: a successful suicide by hanging.)[117] Journalists and their readers became obsessed with diagnosing the circumstances of that overdose: was it a suicide attempt? And, if so, was Dinoire mentally stable enough to withstand the psychological trauma of the face transplant? Or, at a more basic level, was she mentally stable at all?

To her doctors, these were important but already answered questions. As Lengelé et al. insisted, "[t]he patient who received the first face allograft, whose real circumstances before she was bitten by her dog actually remain unknown to anyone, was attended to over a period of 6 months by three different teams of psychiatrists. All of the experts consulted concluded unanimously that she was in a position to understand the issues of the innovative treatment that was being proposed to her and, subsequently, to deal with the constraints of the adjuvant treatment to which she was to be subjected." Speaking to concerns that the doctors were being opportunistic with a patient who happened to be in the right place at the right time, Dinoire's team insisted that she had been carefully—extra carefully—screened, as all potential transplant recipients are, particularly in the case of scarce resources. Not only did Dinoire need a face, she was well prepared to receive one. And, her team insisted, the psychiatrists were right: "Every day that goes by confirms that their analysis was well founded."[118]

These responses highlight the extent to which not just Dinoire's surgeons but she herself were subjected to questioning over her participation in the procedure. Prior to the revelation of Dinoire's life history, reporters framed the surgery in cautious but excited terms, citing it as a major medical breakthrough with some ethical considerations yet to be worked out around risks and benefits. The unnamed recipient remained a sympathetic victim being granted an unprecedented opportunity. As details emerged about the process, Dinoire's surgeons and their methods were subjected to serious inquiry, though the procedure itself continued to be described in transplant rather than cosmetic terms. It was only with the discovery of Dinoire's identity and the subsequent questioning of her mental state, that the cosmetic frame became more consistently deployed, thereby reframing the surgery itself as elective, aesthetic, and unnecessary. The therapeutic benefits were elided or subordinated to a bioethical narrative that rendered the advantages not worth the risk. As Dinoire's identity got written into the story, the possibility of the surgery as a therapeutic intervention slowly got written out.

There are, as I've outlined, differences in the way different constituents discuss the surgery and its ethical frameworks. Surgeons today by and large emphasize the research and record of success, making clear that the technical aspects of both the procedure and the subsequent drug regimes have strong precedents that eliminate a lot of the uncertainty. They also, interestingly, pay careful attention to quality-of-life issues when evaluating the status of the surgery and its necessity in the lives of recipients. Bioethicists are more concerned with the unknowns, particularly around patient reactions to their new faces and their ability to adhere to immunosuppressant regimes. And the popular press tells a story, a story of medical audacity and patient karma, of reaching new frontiers and evaluating how we got here. The press puts the surgery and all those involved on trial. But all three constituents share at least one major source of discomfort about this procedure and what it entails, though they navigate it in different ways. Each of them has to handle that which is being transplanted, and all it implies. Each of them has to find a way to deal with the fundamental instability of identity and its intimate connection to the face. To the face of each individual.

CONCLUSION

For some, there is still a visceral reaction to facial-allograft surgery and what it actually entails. For others, many others, the depth and vehemence of the debates around facial transplantation are but a distant and slightly confusing memory: what exactly was all the fuss about again, when these surgeries are so self-evidently beneficial and life changing? And yet, as I have documented, the controversies were intense and predicated on multiple concerns, anxieties, and unknowns, both stated and under the surface. Time has answered some of the concerns, particularly those in the realm of the unknowability of certain psychological and physical outcomes. The screening techniques have proved robust, and the overwhelming majority of patients have adhered to their drug regimes and coped well with the psychological impact of seeing a foreign face in the mirror. For these patients, the disadvantages of the surgery are dwarfed by that which they have gained, or regained: their lives.

It seems like it should always have been an obvious calculus: the advantages of this procedure so clearly outweigh the risks. But it wasn't always so clear. For some, it still isn't; if you take quality of life out of the equation, the balance is weighted very differently. And many of the bioethicists engaged in the debates around the surgery made precisely that move, considering the risks of immunosuppressant regimes and the psychological destabilization of having the face of another against the benefits of a supposedly elective surgery. According to my survey of the bioethical literature, they decided, based on these considerations, the surgery wasn't worth it. The risks were too high, and the unknowns—particularly around the long-term effects of the surgery—were too great. Unsurprisingly, surgeons and medical practitioners in favor of the surgery were more sensitive to the quality-of-life and therapeutic advantages conferred by the surgery and were careful to include these issues in their discussions, even as they framed those discussions in a medicalized framework. In so doing they were able to highlight the extent to which the facial-allograft procedure had ample technical and medical precedent; the unknowns were limited and hardly amounted to sufficient reason to delay or deny needy patients this obviously life-changing surgery.

IRBs and ethical boards ultimately agreed with the latter point of view, and the first surgery was performed in 2005. As the vibrant and at times highly aggressive public response illustrates, there were still many who harbored deep suspicion of the surgery. These concerns were—and still are—often framed in terms of the long-term risks of the procedure, particularly around the psychological implications of the surgery. As I have shown, however, the more significant psychological concerns lay in the fear of identity transfer and the process of creating these unclassifiable human hybrids, these unprecedented cyborg monsters. Drawing heavily from the narratives established by science fiction and Hollywood film, objections to the surgery were situated in fears about what exactly was being produced and what it would mean for the nature of identity, individuality, and, ultimately, the future of humanity. The doctors, framed as arrogant and self-serving, were put on trial.

If that wasn't enough, these objections were undergirded by the nagging, if often unstated, feeling that the surgery was, if not unnecessary, then in some way cheating. Cheating people of the ability to read the body's history and experience accurately but, more importantly, cheating the fates that had decreed this to be the lot of some, which they ought to bear bravely. Linked to a long history of distrust in cosmetic interventions, this line of objection was couched in the language of risks and benefits, so enabled by a misrecognition of the surgery as cosmetic and elective. The patient, framed as unstable and immoral, was put on trial.

For a while. The patient and the doctors were put on trial for a while. And while echoes of the debate continued to sound for a number of years following the first surgeries, they were and are faint indeed. Partly, of course, the novelty wore off. The surgeries weren't the same kind of clickbait anymore. Partly, subsequent teams had a more careful PR strategy, offering less media access to the individuals and carefully framing the discussion through press kits focusing on the medicine rather than the bioethics. And, partly, the patients were so evidently helped by the procedures and were so clearly managing to navigate the psychological and medical challenges as to render those objections somewhat moot. But, most importantly, the contexts for the surgeries themselves changed. As I discuss in the next chapter, the military entered the discussion in a very public way by funding not only research into the procedure but also some of the surgeries themselves, in order to apply the techniques to disfigured vets injured in the line of service. These vets, the failed sacrifices of ritual national unity activity (which is to say, battle), need to be at least partly rehabilitated in order to maintain the social order. When injured to the point of nonfunctionality, they challenge the collective sense of well-being established by civic cohesion; they are at the limits of humanity, a very uncomfortable place for them to occupy indeed. Uncomfortable for everyone. Restoring them, through face transplants or other means, is not only not objectionable, not only not cheating, not only not elective. It is a fundamental social necessity.

The figure of the injured veteran does a great deal of rhetorical and ideological work to reframe the surgery as therapeutic and necessary rather than cosmetic and elective. So too does the marketing and mediatizing of the procedure, which is highly accessible in terms of availability and legibility. As we shall see in the next chapter, subsequent patients' media exposure was carefully managed around the familiar makeover format to provoke a sympathetic response that overruled visceral yuck reactions and bioethical objections. Explanations of the surgery are easily found and easily digested in a number of visual, textual, and other mediated forms. These explanations reveal the procedure itself to be a careful, painstaking, and often tedious process with profound—but hardly monstrous—results. The countless before-and-after photographs and interviews with the recipients, who are all shown to be grateful for the procedure and thriving in their new lives, detract considerably from the uncertainty about who or what they would be like. They may be cyborgs (aren't we all?), but they aren't monsters. They are yet more participants in a game we know well: makeover reality television. They are yet more happy and surprised recipients of the big reveal. Mediating the surgery to such a degree—a process motivated largely by funding and institutional competition—shifts it from the realm of science fiction to the realm of science. It's not scary anymore. And, in keeping with the conventions of makeover reality television, it's not radical anymore.

CHAPTER 5

A Very Special Makeover:
Face Transplants on Television

INTRODUCTION

What kind of public figure is the face-transplant recipient? If, as Graeme Turner has laid out, a celebrity is someone who generates public interest as much if not more for her private life than her public achievements, what sense can we make of someone whose fame emerges entirely because of her private life?[1] (Because what can be more private, more personal—even as it is that part of ourselves that is most publicly presented to the world—than the loss of face)? A face-transplant recipient is a very particular kind of famous person: one whose (new) face is well known, plastered—usually briefly—across all major media outlets. Someone who is famous entirely for his face, and entirely because that face was once—but is no more—a lack. This cliché, the cliché of the ubiquitous face, takes on particular resonance here, because it is the face itself that is the celebrity, or the cause of it. Face-transplant patients are famous (kind of) by virtue of what has been done to them, both in their loss of face and its remediation. The real celebrities, though, are the ones who make the change happen, the ones responsible for the transformation, the makeover experts. The doctors. The face transplant is a very special kind of makeover, a particular version of the realignment of the nonrepresentative outer core; while the new face cannot be an accurate representation of the inner self, it is surely better than nothing. A face-transplant recipient is not, as in the classic makeover narrative, becoming on the outside the person she always was within. Rather, she has (once again) become a person at all. So some say. Others, as we have seen, disagree.

In this chapter, I'll explore the narratives presented in the television portrayals of face-transplant recipients, at least the ones—a surprising number—

whose identities are known. I will show that the representations of the surgeries and their aftermaths borrow heavily from makeover formats, drawing on elements of the medical documentary but with particular focus on cosmetic-surgery reality television. While there are important points of divergence, which I'll discuss, there is strong resonance between the fundamentally conservative genre of makeover television and facial-allograft stories, which serve to tame this potentially radical bodily intervention into traditional models of gender, identity, and adherence to normalized standards of appearance and self-presentation.

In a way, the reality makeover format is an obvious structure in which to present and package face-transplant recipients. Many of the same elements are present: a life hampered by problematic or nonrepresentational appearance, the opportunity to rebuild fractured relationships and life goals, and, indeed, to refresh and recharge them entirely. There is the opportunity to dwell on what the recipients deserve and to contrast it—rather sharply—with what they have. There are charismatic (or sometimes charismatic—the doctors are a work-in-progress on that front, and they do improve) constants between recipients or episodes in the form of makeover specialists, be they hosts, personal trainers, stylists, or (and) doctors. There are stunning, even magical, before and after photographs and images, which adhere strongly to the logic of the big reveal. And which adhere strongly to the narratives of self-packaging and self-transformation on which these formats fundamentally rest. And, of course, there are the people themselves. They have been made over. They have been, from an external perspective, transformed.

But the new face of the FAT recipient isn't perfect. It isn't seamless. And that matters. The traditional makeover hides or disappears scars, making the trauma of the makeover and that which inspired it invisible. The new faces and bodies that are (almost always) proudly displayed during the highly staged big reveal elide the labor that came before.[2] The elision of trauma is, I suspect, part of both the pleasure in, and the reason for, the showcasing of the process. Even if we can't see in the final product how messy it is to become (more) perfect, we know it to be true. The pain has been performed for us. The results, in the neoliberal logics that many have argued govern the makeover genre, have been *earned*.[3]

With face-transplant recipients, the process is visible in the product. As impressive as the surgery is, the scars are not to be eradicated. Even as the surgeries intervene into the effects of past experience, replacing what was lost, they do not seek to erase history. Face transplants recapture a condition that once was: having a face. But they do not bring back the precise position of the faced past, instead creating something entirely new. Something entirely imperfect, with the experience of time and the labors of the process clearly marked. Which, among other factors we will discuss, helps makes the process more palatable for us. (Because, of course, it is all about us. At least in our own eyes.) The visible trace of trauma helps make it OK. Helps, and maybe even solves, the question of how to categorize the face-transplant surgery as either elective or medically necessary. Through the makeover narrative (and, later, the backing of the military), face-transplant surgery becomes a straightforward (if miraculous) solution to a problem. The problem is unacceptable appearance and the inability to adequately perform as a consuming subject in the world. The solution is to fix it. The solution is to make trauma—the trauma of domestic abuse, of military injury—into the experience of the everyday, just another part of the journey of self-actualization. In this way, the radical potentiality of the face-transplant surgery and the hybridity it suggests is neutralized instead to placate a deeply conservative set of narratives and standards about how to be in the world. And, at the same time, makes these kinds of traumas just another aspect of what it is to be in the world.

MEDICAL/MAKEOVER

Let's get just a wee bit technical here, considering which aspects of makeover television and medical documentary speak to the coverage of FAT recipients. Some of the elements of these two genres, makeover reality television and medical documentary, will be broadly familiar and even, as in the case of "the Treeman" Dede Koswara or chronicles of splitting conjoined twins, somewhat overlapping.[4] Other elements will be a little subtler. Some won't apply, not exactly. But some, particularly the framing, pacing, and staging of the big reveal and its close companion, the before-and-after images and logics, run across all such genres.

Tania Lewis offers a succinct summary of the personal-makeover television format, outlining the rarely deviated-from stages in transforming an unacceptable specimen into a much-improved and more easily presentable final product.[5] The process begins with the surprise visit by the makeover experts, in which the participant is informed that she (usually she, but not always) has been volun*told* for the experience by her friends or family out of their concern for her well-being.[6] Her perceived deficiencies and presentational transgressions are then catalogued in painful and televised detail through accrued documentary footage, alongside her reaction to this public litany of complaints. Based on their expertise, the hosts or experts tailor a transformational program for her to follow both in the moment and going forward, the results of which are unveiled during the big reveal. The participant premieres her new self in a highly choreographed moment that underscores the fundamentally positive—and highly profound—nature of the rebirth and renewal of self and identity.

The process is often a painful one, emotionally and sometimes even physically, depending on what is being made over.[7] Even those shows that seek to highlight and celebrate positive attributes (rather than calling upon participants to change themselves entirely) do so at the explicit expense of the so-called negative ones, which they are careful to delineate in detail.[8] The trauma of the transformation—captured in closely choreographed sequence—is, in many ways, precisely the point and pleasure of the production. It's also the wages by which participants earn the right to their transformation. The scars—be they external or internal—are deliberately hidden in the final product. Scars are indexical of experience and the passage of time; time is that which is to be erased. The new self is meant to be reborn whole, devoid of any overtones of monstrosity or abnormality (or anything else that would compromise mainstream commercial appeal of both the show and cosmetic surgery more generally), appearing to seamlessly embrace her new manifestation in the world.[9]

But that would be too easy.

If they aren't born with it, people need to work for their perfection. The pain of the process, the ritual humiliation, is the labor that participants trade for the expertise of the hosts. And for the right to access the benefits they now deserve as more appropriate consumer-capitalist subjects. It's a market exchange, in its way, in which humiliation and other forms of emotional trial are offered for skilled advice, audience pleasure, and, in the end, the right to be reborn.

Face-transplant recipients are not unwitting or reluctant participants, at least not at the point at which we encounter them televisually. They are not subject to a list of deficiencies for which the makeover is the solution. There is only one significant, glaring, and impossible lack: the lack of a face. Well, perhaps that's not the only one. While the coverage in general is careful to avoid outright blaming the FAT recipients (the media learned its lesson after Isabelle Dinoire, sort of), their lives before the incidents that caused their face loss are also recounted in highly edited detail. As we'll see, the narratives are carefully constructed to highlight their former attractiveness of person and appearance and the happiness they accrued from their lives and (indirectly) from their attractiveness. And how that all changed, and how they, confronted with their new reality, changed. So here, too, there is a deficiency that is to be solved. Here, too, there is to be transformation.

The before portion of the FAT makeover format is somewhat different than the traditional version. There is more emphasis on the positive than the negative, and less emphasis on the ways in which change is needed to call into being the true (and hidden) essence of who the participant truly is.[10] And there is an additional aspect to the before portion, which is more like a during portion, in which the life of the participant as faceless is displayed and discussed. This aspect has to be particularly carefully negotiated, given the visuals of the situation and the need to seem respectful of the patient. It's also not fly-on-the-wall footage; it couldn't be, and it needn't be: the faceless and future newly faced are documented in detail. There is a second half to the during, which is the most direct makeover portion, or the most significant part of it, namely, the operation itself. The medical context, or the more explicitly medical context, makes the makeover narrative fit rather poorly on this piece, and here the format borrows more from the conventions of the medical documentary. Or, perhaps, it combines the two, turning the surgeon into the makeover expert and host, the constant from episode to episode even as the participants themselves change.

The after portion, of course, bears the strongest resemblance to the makeover (and, indeed, the medical documentary), including but not limited to the drama of the big reveal. But here too there are important points of divergence. While the doctor plays the role of the true and consistent celebrity, the FAT recipient isn't a "one and done." His presence lingers and ramifies across programs and formats. He doesn't disappear. He is referred to when discussing subsequent cases. He is revisted for updates. His story goes on.

He's also, in important ways, not done. The person revealed in the moment of the face-transplant big reveal is far from a finished product, unlike the way that the traditional makeover recipient is presented. Part of the reason is straight-up technical: the face itself is, immediately following the surgery (or whenever the patient reveals himself) still healing. Still changing. And that change happens at a rapid, and often highly public pace. In the most literal of ways, the (fading) scars are very much on display. The passage of time is visible and present, and it matters.

The trauma through which the FAT participant purchases the right to the makeover is integral to the narrative itself, and that trauma doesn't end with the new face. The solution is itself resonant with the problem; the participant looks better, much, much better, but still not seamless. Still not normal. He, unlike the supposedly reborn traditional makeover participant, is a work in progress. That's part of the reason why the story is ongoing, why these recipients don't disappear from view immediately following their initial transformation. But it's only part of the reason, and the situation isn't so straightforward, as we've seen with Isabelle Dinoire. Facelessness may not in and of itself constitute legitimate labor by which to earn rebirth, at least according to some narratives. The participants have to deserve it. And they have to continue to deserve it even after they have their new faces, which, after all, are works in progress. And, after all, these faces can still technically (and medically) be lost.

Some of the equation is (or is framed as) medical and technical: if recipients do not treat their new faces and (old) bodies appropriately, the former may be rejected, in which case the new faces can literally slough off.[11] Built into the story is a handy little feedback loop for critique and judgment: smoking, drinking, and otherwise "immoral" behavior can carry very real consequences for the ongoing transformation.[12] But some of the equation is more subtle: recipients' trauma is framed as ennobling to provide the character development and self-auditing so necessary to the makeover format. For those not ennobled for their suffering, for those who haven't undergone change, for those whose trauma—so clearly on display—has not produced a better and more appropriate subject, a makeover of such dramatic proportions would be undeserved. Luckily, according to the television coverage, no such person has yet surfaced. At least not before the surgery. Afterward, as we'll see, may be a whole other story.

Afterward is a complicated time in the space of traditional makeover formats and medical documentaries. There isn't, in the case of the former, a whole lot of it. The portrayals end on the triumphant reveal and (often) return their subjects to regular life to great acclaim and approbation. We don't—we can't—know much about what happens the next day, the next week, the next year. We can't learn the mundane and quotidian (and possibly terrible and traumatic) details that might shatter the fairy tale of happily ever after. This is particularly true in the case of cosmetic surgery and other forms of radical intervention, in which participants may end up worse (emotionally and sometimes physically) than before.

It's of course hard to know all the details of the afterward given nondisclosure agreements and confidentiality clauses, but we sometimes have the evidence of our eyes and access to other versions of the story. Participants on *The Biggest Loser*, for example, often gain the weight back, and more, as we know from various follow-up interviews. The transformation is short-lived, and the relapses can carry intense emotional resonance of personal failure. The self-improvement structure of the show places responsibility for weight gain (and loss) squarely on the (shrinking) shoulders of the contestants, who are taught to understand obesity as a moral lack under their control. In the absence of tight food surveillance, collective guilt and pressure, and rigid exercise oversight, daily life and food control can be an overwhelming challenge. Consider season-one winner Ryan Benson, who lost 122 pounds, only to gain 90 of them back after the show ended.[13] Former contestant Kai Hibbard, who gained back two-thirds of the 118 pounds that she had lost, said that she had "a very poor body image" after leaving the show and had developed a number of compulsive and unhealthy attitudes toward food.[14] Combined with the deep sense of loss and responsibility, contestants can feel worse about themselves than they did before the experience. Rather than engendering a transformation, the show may only heighten already present challenges.

Former contestants from the extreme cosmetic-surgery makeover show *The Swan* report similar postshow challenges, finding it difficult to navigate old relationships with their new appearances, and struggling to maintain their physical transformations. Participant Lorrie Arias has been particularly vocal about her life postshow, saying that she is "agoraphobic, on meds, and unable to enjoy life," much of which she blames on the lack of follow-up after the season ended.[15] These are cases of extreme physical transformation, but even the more superficial interventions prove difficult to maintain; in the absence of professional makeup artists and stylists, some guests on *What Not to Wear* spend most days looking exactly as they did before. They may have been transformed (by others), but then they transformed themselves back.

Face-transplant recipients may also transform back . . . if something goes terribly, terribly wrong. But, unlike the "happily ever after" message of traditional makeover shows, the possibility of failure (and the celebration of success) is precisely where a lot of the dramatic tension and narrative potential lie in face-transplant stories. The chance that the face may be rejected is a very real one, which not only lends excitement and possible pathos to the reports (will the operation be a success? will the recipient keep her face?) but also ensures follow-up coverage (stay tuned!). Afterward, in the structure of the face-transplant narrative, matters very much. And part of the reason it matters, and part of the way that is stays interesting, is because the recipients can (much like in traditional makeovers) play some role in how it all turns out. Not a huge one—medicine and technology bear most of the responsibility for the long-terms success and acceptance of the transplant (or so the story goes)—but enough of one to allow for enjoyable surveillance of the recipients' personal habits.

The big reveal, in the face-transplant narrative, is just the beginning.

That's one way in which the story is different, and it's an important one. It allows the recipients to linger as significant figures, celebrities even, at least for a little while. Some recipients extend their platforms to advocate for related causes, including organ donation, medical experimentation, animal-rights awareness, and domestic-violence survivors.[16] There are other differences, including the way that internal change is conceptualized in conjunction with the changes taking place on the surface: for face-transplant recipients, the personal transformation occurs while living without a face, and the external makeover is the reward for that experience. In traditional makeover narratives, it is the external intervention that either precipitates the internal change or makes manifest the changes that were always going to happen. We'll see how this works in face-transplant stories in just a bit, but first we're going to explore a more significant structural difference, namely the deployment of medical documentary techniques in conjunction with makeover formats.

José van Dijck, in her detailed analysis of the changing nature of medical films in the twentieth century, points out the emergence of the doctor as the hero and subject of the story, with the patient being a faceless object on which the doctor exercises his expertise.[17] Doctors certainly play an important and even heroic role in face-transplant narratives, offering consistency across stories much like the expert hosts in traditional makeover formats. But the patient is far from faceless. The face of the patient is, indeed, the point. Or one of the points. Equally important, perhaps, is the presentation of the faceless patient as a patient at all. That isn't, in fact, obvious.

The patienthood of the face-transplant recipient needs to be constructed. As we've seen across this book so far, the resistance of the surgery to categorization—and the corresponding resistance of the faceless to categorization—makes this a process and a highly individual one at that. It's a process at which television and film (unlike newspapers and scholarly articles) are well situated to succeed. The starting assumption of face-transplant stories is that there exists, in David L. Clark and Catherine Myser's wonderfully evocative term, a state of being that is "corporeally proper," and facelessness violates that state.[18] The transplant procedure is one that restores the body to its proper position through the aid of scientific technology and (heroic) medical expertise. In short, there is a problem (how it is categorized doesn't matter), and the doctor can fix it. And return it to the normal order of things. This is a radical disruption, in its way, of the traditional medical narrative in which the bodily crisis is the precipitating event. Here, while attention is paid to the original injuries and the lifesaving procedures that keep the injured alive, faceless but alive (such as it is), the medical saviors of life play a significantly reduced role compared to the saviors of the self.

Clark and Myser, along with the work of van Dijck, explore medical documentaries that chronicle the separation of conjoined twins as an inevitable and necessary process.[19] Margrit Shildrick's work traces the history of conjoinment in conjunction with medicine, recounting the debates around whether such twins are a singular or plural entity, and what the stakes for these approaches are for autonomy and self-identity.[20] While their studies raise the possibility of radical ethical challenges to the necessity assumption, they all ultimately find the medical documentary genre to be a fundamentally conservative one that reinforces traditional ideas about what constitutes the construction and policing of appropriate bodies.[21] Face-transplant makeover stories share the logic that a face transplant is the ideal medical outcome for a transgressive state of being. Unlike the traditional makeover format, in which the faces and bodies of the participants are problematic but possible, the corporeality of those in medical documentaries cannot be sustained. Conjoined twins must be separated (regardless of the potential medical risks and long-term effects) and facelessness must be eliminated.[22]

A new face, like an independent body, is not a happy ending but a new beginning for a life put on hold.

The comparison between facelessness and conjoined twins is an apt one, with both offering transgressive possibilities for the nature of singular embodiment, the connection between body and identity, and the links between self and other.[23] In both cases, the radical potential is tamed through deeply conservative depictions of medical figures and medical interventions, which, when combined with the rigid gender politics of the makeover story, render this process radical in its newness, and in no other way. The presentation of face transplants under the makeover and medical rubrics reinforces, in its way, the stability of identity rather than its fluidity.

As Scott McQuire has noted (following John Tagg), proof that "I" am someone worth photographing or filming constitutes a peculiarly modern proof that "I" exist as a unique individual.[24] The public personas of face-transplant recipients serve as a kind of conservative testimony to place them among all other equally unique individuals, who are that and just that. In this case, the uniqueness isn't in question. The individuality, the corporeal individuality, might have been (given the possible multiplicity of identities and personhoods), but placed alongside all others, face-transplant recipients become yet another example of highly choreographed and ultimately unsurprising change. Their celebrity, such as it is, lies in the ways that they are like everyone else who have done (or wish to do) the same thing—not transplanting faces, not exactly, but making themselves over. Medical documentaries share with makeover television their insistence on a normative gendered body as the ideal outcome. Makeover television, according to Katherine Sender, falls squarely in the category of so-called women's culture.[25] Medical documentaries, with their emphasis on heroic expertise and gory details, play to a masculinist sensibility. Face-transplant stories, with their triumphalist narratives of transformation through elite and expensive expertise and technology, manage, in deeply conservative ways, to speak across gender lines as a form of edutainment.

In the concluding chapter of this book, I'll spend some time considering what a radical understanding of the face transplant might look like. Now I want to turn to the stories themselves, thinking about how they work and what they are actually saying. Let's (finally) meet some of the other face-transplant recipients. Let's also think about how things changed after Isabelle Dinoire. And about what things didn't change.

MADEOVER

Here are the basic structures repeated across the television reports and specials on American face-transplant recipients:

1. The big reveal of the new face and people's reaction to it.
2. The doctor as hero/expert who provides consistency across patient narratives.
3. A strong emphasis on new social possibilities postsurgery, illustrated by a scene of the recipient in a public place. This is contrasted with the horrors the recipient experienced in public prior to the surgery.
4. The highlighting of personal growth and development postdisfigurement but prior to the transplant, usually with language around how the recipient wouldn't go back to life predisfigurement.
5. Ongoing engagement with the patients as their faces heal and swelling decreases, until the point at which their faces reach their assumed final state.
6. Gratitude. Lots and lots of gratitude.
7. The cheerfulness of the recipient: we like our victim-heroes happy and grateful.

There are different kinds of makeover shows operating under their own internal logics. The face-transplant makeover borrows from various formats;

there is an early emphasis on the loss of what once was, like shows that allow for the reversal of time or the recapture of a body or face that the characters once had. With a twist: the return here is to the condition of having a face, but the face itself is entirely new. Well, sort of. There is some narrative attention paid to the patients' claims that having any face at all is more familiar than having no face. Borrowing from medical documentaries, we have the miraculous intervention of science and expertise in the form of the operation, with the doctors as the heroes and the participants as patients. And then we have the constant across genres, the big reveal. But, finally, the face-transplant makeover transitions into an ongoing present, which has no real analogue in the other formats. In the face-transplant makeover, the afterward continues as the scars fade and the intervention progresses. With face transplants, the transformation proceeds alongside the very real possibility of reversal and failure.

Now let me show you.

In a way, it makes sense to go through the characters chronologically, starting with the first face transplant in the United States, Connie Culp (December 2008), and moving through James Maki (April 2009), Dallas Weins (March 2011), Mitch Hunter (April 2011), Charla Nash (May 2011), Richard Norris (March 2012), and ending with Carmen Tarleton, who had her surgery in January 2013. Except that self-identification and participation with the media did not proceed in the same order as the surgeries, partly because not all the identification was exactly voluntary, and some of it was dependent on other participants and their experiences. To put it baldly, Connie Culp was scared by the reaction to Isabelle Dinoire. She didn't want to be put on trial by the public. It took the significantly gentler media response to James Maki, who had his surgery a number of months after Culp, to persuade her of the advantages of public identification. But Maki had no such precedent. He also didn't entirely have a choice.

The story of Joseph Helfgot and James Maki highlights the challenges of anonymity in the hypermediated environment of these surgical interventions and innovations. Helfgot's family donated his face to James Maki for the second partial face transplant in the United States, performed in 2009 at the Brigham and Women's Hospital in Boston, Massachusetts. Media attention to face transplantation was perhaps at its height, and the press reported in detail on the surgery and its medical team. One of the lead reporters on the case, Kay Lazar of the *Boston Globe*, read an e-mail comment from a reader who noted that his rabbi had cited donations of the face as an example of the ultimate *mitzvah* or good deed, one that can never be repaid. The rabbi commented that a member of the congregation had recently performed precisely this mitzvah. From this lead, Lazar was able to scan death notices to track down the name of the donor and interview his family. The recipient, James Maki, quickly discovered the details of his donor, including what his face used to look like. As did his support personnel and related medical staff. While technically Maki did not have to come forward, his identity was already an open secret. Maki, excited about his new life and renewed ability to eat, speak, have teeth, taste, and soon go out in public, was happy to oblige the interested public with access to him and his story. And an amazing story it was, both for his renewed capabilities and ones that he never had in the first place, including the ability to grow a beard and a recurring case of rosacea that he received from his donor. His decision to go public was likely eased by a visit he received in the hospital from Isabelle Dinoire, whose almost invisible scar lines (trauma disappeared!) and ease of international travel translated to optimism for Maki's future.[26]

Not coincidentally, Connie Culp, the recipient of the first full face transplant in the United States, decided to go public less than three weeks after the Helfgot story broke. These two stories doubtless helped drive one another as Culp was interviewed on such high profile programs as *Good Morning America*.[27] The characters involved (including doctors, donor families, and recipients), the media, and the interested public all combined to keep the transplant in the headlines and keep the debate alive. But already by 2009 the urgency of the bioethical discussion had faded significantly. Even though James Maki had received his injuries by passing out on train tracks with an aneurism while seriously under the influence of the drugs to which he was addicted, he was not subjected to nearly the same scrutiny and moral judgment as Dinoire.[28] He was not put on trial in the same way. Nor were the doctors, nor was the procedure itself. In only three short years, things had certainly changed.

One of the most important things that changed was the way that recipients and their stories were presented. While James Maki could, according to a particular kind of narrative, be blamed for the loss of his face, so too could he be praised for his postinjury rehabilitation. (Though his status as a veteran provided him some leeway. That matters too, as we'll see.) Time, the time to grow and develop, learn and transform, and, indeed, redeem himself, provided the conditions of possibility for Maki to earn his transformation. Dinoire had no such time, and no such redemption. In a play on the traditional chronology of the makeover that is entirely consonant with its message, Maki transformed himself on the inside. His transplant allowed his face to catch up.

Maki made his first public appearance in a brief video interview that accompanied the article in the *Boston Globe* giving details of Maki's life and his surgery. Entitled "Face Transplant Patient Speaks Out," the video shows Maki's new (and still swollen and deeply scarred face), which, he notes, will not cause people to scream as his old appearance did. With his new ability to go out in public safely, Maki wishes to work with vets and do something positive in the world.[29] There is no doubt that this recovering addict and vet has learned from his past. There is also no doubt that he has suffered for his sins. He's earned his new face, and he's going to continue to deserve it through his future actions. Indeed, Maki becomes a strong advocate for the surgery, using his experiences and brief celebrity to speak for others.[30] Interestingly, Maki's doctor, the soon to be almost ubiquitous Bohdan Pomahač, does not appear in this first brief report. He does, however, have a starring role in subsequent presentations, including the *Good Morning America* exclusive (face transplants became kind of *GMA*'s thing) that in many ways set the stage for what was to come.[31]

The *GMA* story opens with a discussion of the history of transplant surgery in the United States, situating the narrative around the very hospital (Brigham and Women's in Boston) where Maki was a patient. There is, in what was to become a repeated trope, reference to the two who came before, Isabelle Dinoire and Connie Culp, and pictures of their new faces. The camera then pans to James Maki, but, before we hear even a single word from him, we turn to the doctor, Bohdan Pomahač, who speaks first. We then learn a little about Maki—his fun-loving and mischievous character as an (adopted) child, his service in Vietnam that left him a heroin addict, his years of challenge that resulted in late-night wanderings. A sudden brain aneurism caused a fall on the subway tracks in Boston, leaving him with burns severe enough to limit his ability to eat, speak, and appear in public. Diane Sawyer dwells on the cruelty of passersby and then pivots to information about Maki's donor, ultimately showing side-by-side pictures of the two.

As Sawyer discusses the donor, she asks Maki about the biggest changes he saw when looking in a mirror. Maki answers, "I didn't think there was any big change. It was that good. It was like I'm looking in a mirror and that's what I see. I see my old self." Maki has transformed back to himself through overcoming addiction and other challenges. He, in his eyes, doesn't look like a new person. He looks like the person he always was. The person he always could become. Maki's new face has become seamlessly integrated into his identity; the scars, to him, are already invisible. In a way, the new face is invisible to Maki too.[32] The discussion of the donor solicits a strong (and expected, and necessary) statement of appreciation from Maki, who says about Susan Helfgot, his donor's wife, that "first and foremost I'm grateful to her and her husband." Maki's cheerfulness, his lack of bitterness and general acceptance of his experiences is emphasized by Sawyer, who notes that "one of Susan's happiest days will be when you have your smile back." Then, and only then, will Maki's transformation be complete, which Maki underscores by saying that he will settle down, maybe go to school. Regardless, he will "do a good job whatever I am doing." He can, now that he has a face. He can do a good job now that "I enjoy life. I'm just starting to live it now." Maki does not add the word "again," and, indeed, he shouldn't: although this is, in its way, a return, it is also a rebirth. This is a new life that Maki is going to live, a new life as a hardworking, cheerful, and grateful, soon-to-be-smiling citizen. One who, as Sawyer points out, will always root for the Red Sox.

Despite having her operation around four months before Maki, Connie Culp's first public appearance was at a Cleveland Clinic press conference one month after Maki's surgery. But Culp quickly made up for lost time, becoming a mainstay on the talk-show circuit, appearing on *Oprah*, *Good Morning America* (three times) and becoming a strong advocate for organ donation. The May 2009 *GMA* spot opened with a clear statement on Sawyer's part as to Culp's particular character that earned her access to the surgery. Sawyer noted that "doctors at the Cleveland Clinic said that she was their first patient because of her forgiving courage, humor, and love of life." In addition to emphasizing Culp's deserving nature, Sawyer gives us some insight into the selection process, which this indeed was. Culp was specifically chosen by the clinic as part of their careful strategy for this pathbreaking and high-profile procedure. While Culp certainly had the right to remain anonymous, should she become a public figure, the clinic wanted to ensure the right kind of return on their investment.

Culp made great media, a friendly and relaxed interviewee who presented a dramatic change. The spot opened with an attractive shot of her before she was shot by her husband at point-blank range, destroying her jaw, nose, nerve endings, eyesight, and facial infrastructure. That's the picture we see next, from multiple angles. Cut to an operating-room image, with masked doctors and instruments, and then a blurred image of someone looking in the mirror and applying makeup. The voiceover tells us that the worst part of the damage, for Culp, was that "she'd frighten young children if she tried to leave the house." That was the most debilitating thing. The picture sharpens, and we see that it is Culp herself, in the first after shot, which shows her putting on face powder and, eventually, lipstick. We transition to the wide shot of Culp sitting down with Sawyer and learn about her enhanced capacities, including—once again—the ability to smell. Culp expresses gratitude to the still-unknown donors, without whom "I wouldn't have a face" and thanks them for "being so thoughtful." The interview is moving and interesting, mostly because of Culp's energy and character, but the big payoff comes right at the end.

In response to Sawyer's prompt about thinking about the past, Culp demonstrates in no uncertain terms her personal growth and development that occurred as a result of her disfigurement. She lays her journey bare, narrating how she learned what was truly important in life when she had to reevaluate the stakes for experience: "You know what's so funny? I was worried about my weight and everything. I'm like that is so foolish. It don't matter what you look like. You're always gonna worry about something. You know, your waist—your weight, your—you know, your hair." She was ennobled by her suffering and brings an uplifting outlook to her life today: "I wanna be positive. I wanna move on. That's what I said. When I woke up from that surgery, it's 2009. Everything's gonna be great from here on out." There is almost an intimation that she was the right candidate for the surgery because she understood that she didn't really need it and, thus, was truly able to appreciate it. It doesn't quite make sense, except that it totally does—the surface may be somewhat superficial, but Culp's depth ensures that she will use these superficial gifts to good purpose. As indeed she does, becoming an advocate for organ donation. Though her journey is not yet complete—Culp, we learn (we are "shocked to learn," as Sawyer says later), is still in love with her husband, the man who made all this necessary. The man who abused and shot her. Culp says, "I still love my husband." She also signals that this (alone) is a hands-off topic: "I can't talk about him, OK?" Even with her new face, Connie Culp still has work to do.[33]

The big reveal here seems kind of rushed. No drumroll, no ta-da moment, no unveiling of the curtain. Just a before, after, and way after sequence of photos, with the pivotal moment being not the surgery but the destroyed face. That's because the news media, alongside Connie Culp, was playing the long game. The *real* big reveal occurred one year later, in the follow-up interview between Culp and Sawyer, which is explicitly advertised as an update about how Culp's face looks now. The spot opens with the now familiar sequence of shots of Culp before the attack, after the attack, and then after the surgery. And then the big reveal of Culp's face now, after the swelling has been reduced and the scars have faded. Culp looks better. And indeed she is, as Sawyer updates. Hearkening back to the prior interview, in which we learned that when Culp was faceless, "children would cower and call her a monster," Sawyer assures viewers that Culp is now, in the best possible way, invisible. Sawyer "took her to a mall." In a "tender blessing, no one looked." The transformation is (almost) complete and is given, during a brief interview with Culp's surgeon, Maria Siemionow, the medical blessing. It is only at the end, when we learn that Culp is no longer in contact with her husband, when she sends a message to abused women "to listen to what he says he is going to do. He will do it," and that "it does not get better," that she is better. That she has finally transformed.[34]

A small postscript: Culp did, in the end, get to thank her donor's family in person. It was a heartwarming Christmas story, in which the donor's family came forward to meet Culp for the first time. Mediated by Diane Sawyer in the television studio.[35]

Dallas Weins, like James Maki, got his transplant at the Brigham and Women's with Bohdan Pomahač leading the surgery. While the prior two American surgeries were extensive, this was the first full transplant in the United States. Weins was painting the side of a church when his forehead touched a high-voltage wire, leading to permanent blindness, and extensive burns that eradicated his lips, nose, and eyebrows. Three years after his injury, he underwent a fifteen-hour operation that restored his speech and sense of smell, though nothing could be done for his lost sight. Two years after his operation, Weins married Jamie Nash, a fellow burn patient whom he met at a burn-survivors support group. In a nice narrative of closure, the wedding took place at the Fort Worth church where Weins was originally burned.

In what has become almost a ritual, Weins appeared on *Good Morning America* to debut his new(ly regained) smile.[36] But the emphasis in this story, the big reveal, is framed through the eyes of Weins's young daughter Scarlett. Part of this is practical; as Weins is blind, he cannot discuss his own personal transformation to the same extent.[37] And part of it is motivated by more abstract concerns: as a man, Weins does not have to be as concerned with his physical appearance. Indeed, as he says in the ABC interview, "I could have lived like it was no problem . . . if I did not have my daughter." This message, the message of Weins's bravery in the face of a terrible condition, is underscored throughout the interview. Even as his new abilities—to smell, to feel— are emphasized, his blindness is unchanged. For most, the voiceover tells us, this would be "a life-shattering prospect." But, "not Dallas."

Why not? Because, as Weins says, "I would not change a thing." The interviewer clarifies: "You wouldn't go back and take it back? Why?" Going back, Weins answers, would mean erasing what has happened since. All the positive changes that have occurred since he lost his face: "Too many good things have happened from it. My family is closer now. I'm a way better person now."

Weins has improved, been transformed, become more noble. The transplant just allowed his face to catch up. The final note reminds us what is really at stake here: "I'm even a better father."

From the outset of the spot, viewers are primed to understand the tragedy of Weins's disfigurement through the context of his fatherhood. All the pictures of him prior to the injury show him with Scarlett, and he is described as a "loving father to his only daughter Scarlett." But before we hear from Weins himself, we hear from his doctor, who is introduced with the familiar overhead shot of the operating room. Pomahač makes clear the transformative stakes of the surgery, reflecting that "it is so profound to see someone who has no face, and then suddenly there is this new person." The voiceover tells us that this is "an exclusive look at the first full transplant in the US." The stage is being set for the big reveal, even as we have already seen Weins's new face. What, then, is going to be unveiled?

Viewers have already seen Weins's new face. Scarlett has not. And then she does, sharing this private "father-daughter reunion" not only with her father but also with the television cameras and with the world. That reunion is "tinged with undying gratitude," as the interviewer JuJu Chang notes, to the donor, who in Weins's words "has given me a new life." Lest we read this as Weins's disavowing the old life that was so very transformative, he continues to remind us of who this is really for: "It's given my daughter a new life." Weins didn't do it for himself. He did it for her, the girl in his life.

Mitch Hunter was also "pretty content with life" without a face, though he acknowledged that "it would give me that one extra step to have a better quality of life" because "I wouldn't get the weird stares."[38] "Weird stares" hold a decidedly different valence than Culp's terrified children screaming about monsters, though he later acknowledged his effect on children as being rather more dramatic, telling a BBC reporter that "I've had kids hide and run behind their mom, scared, and that was kind of hard to cope with." But cope he did, until it became personal: "My friends starting having kids, my brother had a kid, I didn't want kids to be scared of me."

For Hunter, the new face was a nice benefit rather than a lifesaving necessity, as he made quite clear in the BBC interview, underscoring (in response to the required question about gratitude) that his anonymous donor "didn't save my life." Instead, Hunter is careful to clarify, "he changed my life. He saved a lot of other people's lives."[39] He contrasts the face that he received from the donor with other, specifically lifesaving organs that others were lucky enough to get. Others who would have otherwise died without them. Hunter didn't seem to experience his facelessness in the same way as some of the other transplant recipients, or at least he didn't talk about it in similar terms. So we didn't hear quite as much from, or about, Mitch Hunter. At least not through the legacy news media.

The lack of coverage of Hunter's story is rather puzzling, actually. The story of his injury is a great one: compelling, emotional, and full of personal bravery and unfathomable courage in extreme situations. Mitch Hunter, a private in the military, had been a passenger in a car accident that caused a utility pole to fall on top of those in the vehicle. Hunter pushed the ten thousand–volt live wire off the driver and surely saved her life, but in so doing suffered a massive electric shock that caused the loss of his leg and the fingers on one hand. His face was also deeply burned and scarred. While Hunter later recounted that he had no recollection of his action, he credited his decision to military training and instinct in one of his personal YouTube videos.[40]

This story seems made for media. But Hunter's transformation was covered by only two outlets: a BBC documentary and a report of his homecoming on the local news. The local news! That's it![41] It wasn't that people's appetites were sated by face transplants, as witnessed by the numerous stories on subsequent recipients. It wasn't that Hunter himself was media-shy; he willingly participated in the two televised pieces and indeed has an active social media presence in which he talks openly about his surgery. He has released a number of YouTube videos and even answered a very popular Reddit AMA (ask me anything) in which he spoke about deeply personal aspects of his life and his experiences pre- and posttransplant.[42]

Mitch Hunter doesn't make good television. He's not a particularly charismatic interviewee, but even that isn't the issue. It's that his transformation isn't dramatic. Not in the right ways. The appearance piece is impressive. Truly impressive. His "after" images are the most seamless, least scarred, most integrated of all the recipients to date. The lack of scarring, as I've suggested, is part of the problem. The trauma is too hidden, the transformation too complete. Not only on the surface: the internal trauma is also too obscured. We know, from Hunter's own YouTube videos and AMA, that he had experienced great trauma indeed, even considering suicide following the accident.[43] He just didn't tell anyone, or rather, he didn't tell any television reporters. It made the narrative of the makeover too much of a stretch. This already brave, already superhuman man didn't need to transform internally. And he already accepted his disfigurement, not acknowledging the suffering and therefore not being able to be ennobled by it. As only "one extra step" to "a better quality of life," the transformation was too mild. There was simply nothing big enough to reveal.

Charla Nash was next. Same formula: a photograph with a voice-over describing a "loving mother who looked like this" before she was attacked and mauled by a chimpanzee, losing her face and hands, and an image of her afterward. Sort of. Ann Curry describes Charla Nash's "new lease on life" as a lead-in to the story about "the new miracle" (following the earlier miracle of her survival after the chimp attack) of her face and double hand transplant.[44] What we don't see, what we never see, is Nash's face after the attack but before the surgery. She wears a veil. When we do hear from her (even before we hear from Pomahač), she still wears the hat and a veil, at least most of the time. She did once show her face before her FAT on the *Oprah Winfrey Show*, removing her veil in dramatic fashion for the first time in public. In this reverse big reveal, Nash told Winfrey that she deliberately kept tight control of the circulation of her image, noting that she knew she could get millions from the tabloids for pictures of her disfigured face, but she chose to do it in her own time and on her own terms.[45]

We don't quite see her face after her operation either. Not yet. In a drawn-out version of the big reveal, Nash's face is blurred in the *Today* show exclusive, as Nash chose to show her new face to her family (pictured in photos with Nash's face blurred) first. The initial reveal, then, is the family reaction, much like in Weins's story. It's the best *Today* can do in the circumstances, focusing on an interview with Brianna Nash, Charla's daughter, to express gratitude to the donors and claim exclusive access to the story. It's a teaser, another choreographed step in the raising of the curtain. The curtain will soon be fully raised in various contexts, including Nash taking a role in advocating awareness for primate safety with the humane society and starring in her own fundraising video for prosthetic hands after her bilateral hand-transplant surgery (done at the same time as her FAT, in a global first) failed.[46] Nash sends a message of personal bravery and self-empowerment, emphasizing that the more she can do for herself, the better. She is still being rehabilitated and transformed into a visually palatable and self-reliant subject-citizen. A new face was one part of the journey, but hers, as marked by increasing revelation of her appearance, is far from over.

We hear a bit more about Richard Norris, whose transplant in 2012 is notable particularly for how good he looks after what was the most extensive facial transplant to date. Norris lost his face through a gunshot accident, the details of which are glossed over in most of the coverage, likely because he was holding the gun that destroyed his own face. That's not a sympathetic character around which to build a story of deserved redemption, particularly as Norris never cast himself as someone who needed to be redeemed. There are a lot of different and competing stories in the press about Norris, some of which cast his pretransplant life as that of a hermit; others—particularly a highly critical piece in *GQ*—excavate his girlfriend from that period, his job, and his successful adaptation to this way of being in the world.[47]

The *GQ* piece, which is fascinating in its own right for its moralizing and skeptical tone, came out well after the hour-long Ann Curry television special that aired on NBC. Curry acknowledges the accident that caused the disfigurement—of course—but doesn't dwell on it in the same way that reporters dwell on the precipitating events for other recipients.[48] The narrative tension and focus for this program follow some of the traditional makeover guidelines: there's a big reveal, in which Norris's face is unveiled both to himself and to us, the audience. We are privy to Curry's "exclusive interview" following the transplant, as well as Norris's meeting with the donor's family. We encounter the doctor, Eduardo Rodriguez, who performed this particular instantiation of the medical miracle. We learn of Norris's struggles: "he became a recluse" who "had to hide himself, not just from the world, but also from his friends." Norris "sank into a deep depression, even contemplating suicide," turning to the surgery as "a way out" offered by "a bold surgeon" and heroic figure.

But the personal transformation story is somewhat muted. Partly this is a function of time: Norris lived with his disfigurement for fifteen years before the transplant, the longest of any of the recipients to date. He had, it seems, learned to live with it, somewhat. He's also a bit of a tricky subject—a drinker and smoker even after the transplant, activities that could imperil his new face and maybe his life.[49] But, given the extensive nature of the transplant and how good Norris looks afterward (which, given its seamlessness, is part of the problem), the story was too good for Curry to pass up. So she came up with a different narrative, leaning more on the medical frame to provide tension. She didn't have to reach far. She emphasized the risk and who (else) Norris was taking it for, and she followed the money. Quite a lot of it, in fact. Over $13 million to the University of Maryland Medical Center alone.[50]

Norris's transplant, like every American procedure aside from James Maki's, was funded by the American military. But the case of Richard Norris was a bit different: he was explicitly part of a military-funded experimental protocol, the culmination of which was his surgery. The first transplant to be performed at the University of Maryland Shock Trauma Center in Baltimore after a ten-year research project, Norris's surgery held rather more risk than the other procedures, or so Curry presents it.[51] The question here was not how Norris would look afterward, but whether there would be an afterward. Would he live or die? Would his heroism, the heroism he displayed by undergoing the surgery that would grant him a new face and a more comfortable public life, be rewarded?

Curry opens the special by telling us that Richard Norris "would risk everything," which means, she later clarifies, "he could die." And he did it not for himself, not for a chance at a better quality of life, or at least not primarily. He did it "hoping his sacrifice would help him and many others." Specifically, we later learn, his surgery would one day help the "wounded warriors" whose condition motivated the Department of Defense to set up the research grant in which Norris was participating. Norris's doctor, Eduardo Rodriguez, makes "regular progress reports" to the DoD as part of his research grant. The bravery narrative here is situated not in the ways in which Norris lived with his disfigurement. He found a different way to become a better person.

In an interview later in the show, one of the medical staff quotes Norris on the eve of the surgery after he is offered "a last chance to back out." Norris reportedly responded, "no. He said let's go. I'm here. I want to do this. I said this could be the last conversation you and I ever had." But, Norris insisted, he would go forward "to help other people." Curry repeats, "to help other people." This point is again reiterated when Curry interviews Norris directly, asking "why was it so important to risk your life?" Norris pauses, meaningfully, and answers, "if I have to sacrifice my life to help the families get back theirs, it's worth it." Curry clarifies "that your suffering would have had meaning?" "That's right," he answers. And, Curry confirms, "Richard's surgery might in fact make an impact for future patients after some remarkable discoveries." Not just any patients but a particular group: those wounded in service of their country.

Norris does get some of the transformation treatment. Curry performs the familiar ritual of discussing what it is like for Norris when he debuts his new face, telling us that "Richard's new life means new freedoms, like simply going outside without worrying ... now he's just another face in the crowd." She asks him about his first time going out with his new face: "Did people stare at you?" "No." We are told that "in his new life, Richard is already spreading his wings. He is studying information systems technology." And, like other transplant recipients, he speaks of his commitment to the procedure, hoping "to start a foundation to raise awareness about the value of transplant surgery." (According to the *GQ* article, he stops taking his online classes, and the foundation doesn't emerge. But that's not the point. Norris earns the surgery by his very willingness to undergo it. For the sake of others.)

The Norris story was an outing, of sorts. Military involvement in face transplants had hardly been a secret, but with Norris the extent of their funding and interest became clear. The Baltimore facility where Norris got his transplant was, in a way, the military's facility, and Rodriguez, more than Siemionow and Pomahač, was their guy. The other surgeries may have been funded by the military. Rodriguez himself was funded by them as well.

That matters. It changes things. Injured soldiers don't need to earn the right to transformation. They already have.

So had Carmen Tarleton. The most recent of the publicly known face-transplant recipients, she received "the most horrific injury a human being could suffer" at the hands of her ex-husband, Herbert Rodgers, who attacked her with a baseball bat and industrial-strength lye.[52] The chemical burns she suffered eradicated her eyelids, left ear, most of her vision, and her face. It's a terrible and heart-wrenching story, made more poignant by Tarleton's beauty before the attack. And by her charisma and resilience, which is apparent in the many interviews and public appearances she gave prior to her transplant, including a spot on the *Dr. Phil* spinoff *The Doctors* (which comes with an online warning that the images might not be suitable for young audiences. Tarleton did not hide from children, but the programs that featured her thought that perhaps she should).[53]

The story on *The Doctors* was entitled "Carmen's Survival Story." A post-transplant piece on *ESPN* was called "Carmen: A Survivor's Story."[54] The Tarleton narrative was clear. A brave, plucky woman, she gave good TV, even asking to receive a kiss from Travis Stork, the handsome doctor-host of the show. This is notable because of the contrast in their appearances: most guests wouldn't need to ask, and we wouldn't even notice the absolutely standard cheek-to-cheek kiss between interviewer and visitor. But, in Tarleton's case, it's a shocking and rather daring request, granted because her inner character was so powerful that the doctor was (even) willing to touch his lips to her hideous visage. The kiss, rather than undermining the importance of appearance and our awareness of it, only emphasizes it. And the audience cheers. For Tarleton, sure, but really for the doctor who honors her character by himself bravely kissing her missing face.

At this point the basic structure of face-transplant television coverage is familiar: pre-attack photos, postattack images, and then a teaser leading up to the postsurgery reveal. A discussion with her doctor. Gratitude. A testament about the "first time I walked in a grocery store in years and nobody looked at me—wow, I all of a sudden fit in, when I hadn't in quite a while."[55] And I don't want to give short shrift to Tarleton just because she comes last in the narrative, and we now know how the story goes. That shouldn't be taken as a diminishment of her pain, or suffering, or bravery, or resilience. Carmen

Tarleton is an amazing woman. Whose story has a nice twist, one that played particularly well into the story of triumph, change, and transformation of the self. All of which came before the transformation and restoration of her face.

Carmen Tarleton fell in love. Before her transplant, after her attack, Carmen Tarleton fell in love. And she was loved back, by her piano teacher, who was drawn to her prior to her surgery and stayed with her afterward, a man who describes her as "a beautiful person, inside and out. Her spirit, her kindness. She's just a warm, loving human being." (And we admire him, we're supposed to admire him, as we do the doctor who kissed her, for his tremendous ability to see beyond her face.) It's a nice happily ever after, enhanced by Tarleton's new face rather than facilitated by it. There's no doubt, from what we learn about Tarleton, that she had undergone tremendous trauma and demonstrated tremendous personal courage and resilience. And, we learn in the *ESPN* interview, that she will overcome the trauma, internally, as the "emotional scars began to heal." The external trauma remained, as the "physical scars continued."

Tarleton, who despite getting her transplant rather late in the game for it to be called pathbreaking or experimental, garnered quite a bit of media attention. Partly this was due to her compelling personal story, though as we've seen with Mitch Hunter that isn't quite enough. Partly it had to do with her own charisma and charm, which goes a long way in the television world. And partly it had to do with how nicely her story fit into the very particular kind of makeover narrative that emerged around face transplants. To wit: Tarleton was ennobled by her suffering, emphasizing that she "wouldn't go back to her 39 year old body" because "positive events propel us forward to the people we want to be." She specifically addresses the question of domestic violence, telling viewers that "there is a way to overcome it. There is a way not to live in fear. There is a way to find happiness and joy." She had already transformed inside; the transplant allowed her face to catch up as "a great way to move forward with what life has for me now."

Tarleton's transformation is far from seamless; her eyesight remains weak, her hair in many places does not grow, and her face doesn't look quite right. But it's evolving, even as she herself is evolving. Her change, which includes forgiveness of her ex-husband, is tracked through her face. It's an ongoing story, and it's not happily ever after yet. Tarleton underscores that for us, making clear that she wasn't a Pollyanna even in her acceptance of her scarred self: "I had many challenges. I still do." Having a face doesn't equal a perfect life, for Tarleton or anyone else: "Life doesn't get easier, you just get better at it."

And sometimes, life gets harder. Especially for some.

A VERY SPECIAL MAKEOVER

With or without a facial disfigurement, some things remain the same, namely, strongly gendered expectations about the meaning of appearance and the stakes for having an attractive face. As Katherine Sender argues in her comprehensive book *The Makeover: Reality Television and Reflexive Audiences*, makeovers explicitly address crises in gender, resolving them along white, heterosexual, middle-class lines.[56] Certainly both men and women may want to avoid the scrutiny and worse that comes with disfigurement. Certainly men may feel that facelessness is a barrier to finding a romantic partner or navigating social relationships and public interactions. For women, it is fundamentally assumed. Again echoing the logic of the makeover, which transforms participants from un- or underacceptable status as consuming professional and social citizens, the face transplant renders such forms of public participation possible. For Isabelle Dinoire, initial reactions to her new face emphasized, among other things, that "when the scars fade, she could be considered pretty."[57]

Well, that's a relief, since, if she's considered pretty, she might be able to find love again. The stories about Dinoire emphasized that she might be able to find friends or go out in public again as the initial goals, but love was not far behind. Dinoire is a bit of a tricky case because, unlike the rest of the cases we'll examine, she only lived with her initial disfigurement for six months. That brief time frame didn't allow her the space and experience for the necessary long-term trauma (time, and the depth of the emotional scars, matter) that would earn her the right to her new face and new life. According to the transplant makeover logic, she didn't have a true chance to be ennobled by her suffering, and she didn't really get the opportunity to truly understand how transformative the new face would be. That, perhaps, is one of the reasons that Dinoire has to date treated her new face with suspicion, failing to integrate it completely into her identity and sense of self. Dinoire stands as an outlier, according to most reports, in her failure to completely adopt and integrate her new face, which she describes as "half me and half her."[58] In her diary, Dinoire writes, "I cannot forget her. I cannot and I will not. She exists in me."[59] But even hybridity can be attractive, as facially disfigured writer Barbara Robinette Moss said in a widely quoted statement, saying about Dinoire's new face, "I think it's going to be beautiful."[60]

Restoring beauty is a goal that people readily understand. Especially for women.

Globally, there have been about thirty-one surgeries to date (with some uncertainty about what is happening in China). The ratio of male to female recipients is approximately 5:1, with at least one patient's sex/gender remaining entirely unknown (recipient in Ghent, Belgium, in January 2012.)[61] More specifically, there have been only five female recipients worldwide. The first was Isabelle Dinoire, in what was a highly opportunistic and idiosyncratic situation. One was a recent and anonymous case in China. The other three were all in the United States. The number of cases is quite small, but the concentration of female recipients in the United States is striking in comparison with the global arena. Of the eight transplants that have been done in American facilities, five of the recipients have been men and three have been women. That's not a big difference, or, more accurately, that's a small difference between male and female recipients, even given the very limited sample size.

We have to unpack both the international male bias and the more balanced US picture through some contextual framing of the institutional procedure by which recipients are selected. In order to become a candidate for the procedure, potential recipients need to have a strong advocate for their selection among the surgeons. With so much demand and so few viable donor matches, doctors exercise tremendous control over access to the surgery.[62] Unlike whole-organ transplants, which operate under strict protocol and state-wide and national registries and ranking systems, allocation of donor faces is a highly specific process. One of the very few qualified surgeons currently trained to do the procedure must adopt the case of a specific patient and must be on constant alert for a potential match. The surgeon must also be fairly confident of securing funding for this highly expensive and to date still experimental procedure, at least from the perspective of health-care providers.

In the United States, medical-insurance provider Aetna's most recent policy statement "considers face transplantation . . . experimental and investigational because of insufficient evidence of safety and effectiveness in the peer-reviewed published medical literature."[63] While there has been significant clinical support for the procedure even as of this report, Aetna and other private providers are notoriously conservative in their funding allocation. This makes sense: the surgery and its associated medication is very, very expensive. Luckily, the American military (who has quietly and now more publicly been funding both the research and the procedures themselves for many years) has lots and lots of money. And they care an awful lot about being able to alleviate the effects of facial disfigurement. I'll talk more about military involvement below, but first I want to explain the gendered implications of the lack of protocol for patient selection, and why the United States diverges from the general trend.

Men have more advocates, more access, more privilege, and more power than women. That's a bald, highly general, and simplistic statement, but it has very real effects in this specific instance. There are scores of candidates for face transplants; women as well as men suffer from facelessness and near-facelessness. And, indeed, women are particularly vulnerable to injuries caused by abusive male partners. Two of the three female recipients in the United States were disfigured in just this way. The incidence of female disfigurement caused by domestic abuse is horrifying, all too common, and very much muted in the bulk of the coverage. Aside from the space allotted for forgiveness and moving on, which is cast as personal growth and development. Even here, the onus is on the (female) recipient (of abuse, of a new face) to overcome her anger and pain and challenges. To put it another way: one commonality that is not present is a discourse about abuse, even as attacks, including acid attacks, account for a great deal of female facial disfigurements globally.

Outside the United States, almost no women have been selected. And the problem isn't donor availability, at least not entirely. In fact, cross-gendered transplants have been a viable clinical possibility offering "equivalent, anatomical skeletal outcomes to those of sex-matched pairs" for a number of years, according to an article in the *Annals of Plastic Surgery*, the leading journal in the field.[64] However, despite guidelines designed to control for gender biases, studies have shown that significant gender differentials also occur in access to renal and heart transplants, which are robust enough to speak to a general trend.[65] The face-transplant case, small as the sample may be, shows an even more acute bias, but it would, in absence of regulations designed to resist precisely this pattern.

But the situation in the United States is different. Connie Culp, Charla Nash, and Carmen Tarleton comprise a significant minority among the full group of American facial-transplant recipients, which also includes James Maki, Dallas Weins, Mitch Hunter, Richard Norris, and the most recent recipient, a middle-aged man who had remained anonymous until recently, volunteer firefighter Patrick Hardison. It's also worth noting that US recipients are featured in television specials and news reports to a far greater extent than all the other patients, including Isabelle Dinoire. These two pieces—the more balanced distribution of faces across genders in the United States and the greater participation of American recipients in television portrayals—are linked. Women make good subjects for makeover television. Very good and, in many ways, better than men. The genre, as Katherine Sender has argued, was designed with women specifically in mind as both audiences and participants.

Structured fundamentally around restoring women to normative gender behavior and appearances, makeover television is made for women. While there are some exceptions, as Sender discusses, most such shows adhere to a highly heteronormative and gender-conforming set of expectations that serve to reinforce traditional gender roles.[66]

So it's no accident that the one country that most effectively packages and promotes its recipients in a female-friendly format is the one that has the most female patients. It's also no accident that the one country performing the surgery that has no meaningful form of universal government-funded health care is the one that has the most robust industry around the recipients themselves, almost none of whom (in contrast to the international scenario) have remained anonymous (willingly or otherwise).[67] These makeover narratives raise the profile of the hospitals and the surgeons involved in the procedures, so much so that the Brigham and Women's Hospital absorbed the cost of the first transplant that it did, which was over $250,000.[68] In the competitive health-care marketplace of the United States, there are very high stakes to cornering the market on a medical intervention. There are equally high stakes to figuring out how to get it paid for. Publicity helps.

There's a lot to unpack in these stories and reports, and a lot to compare. I have focused on the American patients and American representations because there are so many of them, and so few of the others, which are largely limited to postoperative press conferences. I am particularly interested in the television stories here, though there are also plenty of magazine articles, newspaper stories, memoirs, and biographies. But they tend to follow other conventions, and what I'm interested in here is the combination of makeover and medical documentary as a way to understand how the most radical of bodily interventions has been rendered deeply conservative. Also, of course, I'm interested in what that says about the potentialities of this innovation, which I'll turn to in the final chapter. And (this may be obvious, but it bears saying), I observe how the makeover narrative helped quell the bioethical angst around this intervention, and consider how it ought to be categorized. The makeover story overcomes the resistance of the facial allograft to categorization by showing how it doesn't matter: with enough labor and trauma, and for dramatic enough results, willing participants should be allowed to have a highly risky and potentially dangerous intervention. Particularly (but not only) if they were injured in the line of some duty or other. (Well, one duty, one service—to the country—in particular.) The makeover structure shows not only why it is worth it but also how it is earned. Cosmetic procedure? Whole-organ transplant? Both? Neither?

Who cares . . . as long as an acceptably transformed person comes out the other side?

Well. Some people care very much. For those disfigurements earned while fighting our battles, transformation is not just a matter of social resurrection. At stake is nothing less than national identity.

MILITARY

The US military was always involved in paying for face-transplant research and, ultimately, for the transplants themselves. They care about injuries, and they particularly care about fixing visible reminders of military damage. We all do.

Carolyn Marvin has explored the role of military sacrifice as a unifying ritual in American patriotism, which she argues is a civil religion organized around the flag.[69] Survivors of these sacrificial attempts, then, become sacred objects to whom communal support and assistance is due. The sacrificed, which is to say the soldier killed in the line of duty, performs an important ritual function in uniting members of a national group. The triumphant returning soldier provides an occasion for ritual celebration and memorialization. The injured vet occupies an uncomfortable space between failed sacrifice and damper on celebration. The absent face of an injured veteran can be thought of as a defiled flag whose mending becomes a matter of religious and holy dictum. It's not just a matter of an uncomfortable reminder of failure: it's a deeply visible one. The awkward space that facelessness occupies between physical injury and illness, emotional and mental trauma, debilitating social barrier and ennobling life event is simply less complicated for veterans. Facelessness, in that case, is a problem that needs to be solved. Restoring and rehabilitating veterans to normal appearance and, ostensibly, to normal life is a question not just of assistance but of social necessity. It is nothing less than the reestablishment of social and religious order.

In *Homo Sacer: Sovereign Power and Bare Life*, Giorgio Agamben describes the condition of bare life, in which particular individuals are excluded or abandoned from participating in the political life of the polis, existing in a state of exception that has, in many ways, become the rule.[70] Injured veterans in many ways exemplify the condition of bare life; in the case of those with severe facial disfigurement, face transplants and other interventions represent a way to return them to society. Rather than acting as a means of creating monsters and social and behavioral anomalies, face transplants, when thought of in conjunction with military injuries, become a way to restore the political and social equilibrium for recipients. In conjunction with injured veterans, the face transplant doesn't resist categories: it transcends them. The injection of the military into the facial-allograft conversation fundamentally altered the underlying understanding of the surgery. Even prior to its use on veterans, the surgery became a different kind of object when the military began to fund not just research around it but also the procedure itself. It became a way to restore society rather than undermine and destabilize it.

The US Department of Defense understood that very early in the face-transplant game. The Brigham and Women's Hospital at which Pomahač received a $3.4 million grant from the Armed Forces Institute of Regenerative Medicine, which paid for three of the first four surgeries there, has now procured funding for six additional surgeries, of which one has already been performed. Five patients have been selected for the next stage and await suitable donors.[71] Deeply interested in the possible applications for a wide range of battle wounds, the military recently expanded its program to include experimental research into antirejection medication, with Charla Nash as its first subject. Nash, like the other face-transplant recipients, relied on DoD funding to pay for her surgery, and, like the others aside from Vietnam War veteran James Maki (the lone recipient to date not funded by these grants!) and then-soldier Mitch Hunter, has no official connection to the military. She is now the subject of a study that will try to wean her off immunosuppressant medications, which cost around $10,000 per year, and replace them with cancer drugs that may have fewer long-term life-threatening effects.[72]

The University of Maryland transplant facility has an even longer-term connection to the military. The Office of Naval Research set up a $13 million research program at the University of Maryland Medical Center for surgeon Stephen T. Bartlett to study composite vascularized allografts for facial injuries caused by explosives, which ultimately provided the funding for the Norris transplant. Following Norris's successful surgery, Bartlett publicly credited the military for the Maryland program of research and transplantation, revealing explicitly for the first time the depth of the military's involvement with experimentation in this arena: "We began this research more than 10 years ago when we saw the devastating injuries sustained by soldiers in Iraq and Afghanistan from improvised explosive devices. Now having seen how this surgery has changed Richard's life, we are even more dedicated to researching ways to improve facial transplantation and helping more patients, including military veterans, return to normal lives after undergoing this same surgery." The main surgeon on the Norris procedure, Dr. Eduardo Rodriguez, was more explicit that "the ultimate goal of this project was to treat the wounded warrior . . . to make them appear well, but to also ensure that they function normally."[73] Normal, here, means as part of society, and maybe even as soldiers once again.[74]

But it wasn't just the doctors at the University of Maryland, of course. Dr. Bohdan Pomahač, the face-transplant surgeon at the Brigham and Women's Hospital in Boston, underscored the centrality of the military to the development of the surgery by noting that "[n]one of this would have been possible without the vision and leadership of the Defense Department." Quite literally, in fact: "The military has absolutely played a critical role. There are no other [funding] sources available for clinical research on face-transplant surgery other than them."[75]

For Bartlett, Rodriguez, Pomahač, and presumably the military itself, the benefits of the surgery so clearly outweigh the medical and financial costs as to render the ethical discussion and indeed the question of categorization moot. This matters: it may be a makeover, a kind of a makeover, but it isn't really an optional one. The transformation at stake, for them, isn't one of self-improvement, or even one of having the face catch up to the changes already made inside. It is, instead, making a person whole. Again, or maybe for the first time. It doesn't really matter. Their goals are to allow the injured to live normal lives, and they see the surgery as the best and most effective way to do that. They show little concern for the psychological considerations of doing the surgery, so clearly are they dwarfed by the psychological considerations of not doing the surgery.

The military's interest in, and unequivocal championing of, the surgery may be part of the reason for the declining hysteria and bioethical debates around the procedure. While the military had long supported research into this area, its very public funding of the surgeries themselves provides a powerful body of support for the endeavor for both veterans and civilians. The Department of Defense continues to make strong statements about its commitment to these procedures, going on record that "[t]he Defense Department intends to continue funding immunosuppression research and collaborating with civilian medical centers to perform more face transplant surgeries." And, aware of their monopoly on funding in this area, they intend to keep doing it for everyone, according to Army Col. Barry Martin, chief plastic surgeon at Walter Reed Military Medical Center, "We want to keep making gains and take care of injured folks who do not have real good conventional options . . . we are actively involved in research and science that will help out not only injured service members, but other folks who are victims of trauma."[76]

The military itself is often controversial for some in the United States, but support for those in the service is high, particularly for those injured while on active duty. Their right to a regular appearance, to a normal life, to not have to bear these burdens, is not really up for debate. Placing a disfigurement in the framework of an injury, a battle injury, makes it a very different sort of thing than a disfigurement to be tolerated. Injuries are something to be cured. The question of cosmetics and their ethics does not enter the debate for injured vets. And, by implication, for those who, through their status as experimental subjects, serve to help these vets. And, by implication, anyone else who is a candidate for this same procedure, namely the face transplant. The military involvement helps solve the conundrum of what kind of object the face transplant may be. It's a treatment for an injury. It matters how the injury was received, kind of. But in the end, not as much as one might think. Not anymore. This kind of trauma (like other kinds of trauma, like abusive men attacking women, for example) is a part of life. More than that: a part of the journey toward the best version of the self. The production of the face transplant is a production of the everyday. It's a neutralization of the everyday, the tragic traumas of the everyday.

CONCLUSION

With face-transplant recipients, our interest is only in their private lives and how they have changed. That is where their true achievement lies, and it is bounded; once they stop changing in dramatic ways, we stop caring. But, given the time it takes for a person to adjust to facelessness, and then the time it takes for a new face to adjust, the tenure of the recipient's fame is somewhat longer than the traditional makeover recipient's. As long as it takes for the makeover process: the before, the during, and the afterward, which, unlike traditional happily ever after narratives, remains ongoing. The face-transplant story has the potential to acknowledge that human change and transformation are ongoing, with or without medical intervention. But that potential is neutralized at the outset through the imposition of fairly rigid categories of what constitutes appropriate change and how the visual correlates are earned. These categories are established through a highly ritualized narrative choreography designed to highlight trauma and triumph as the preconditions for personal transformation. As we learn the carefully excerpted personal stories of the recipients, our interest is piqued, and these medical monstrosities are humanized at the outset. Partly, of course, to show that the transplantation (transformation) is deserved. And deserved not only on the basis of the life that the recipient led before face loss—though that, too, sometimes (and sometimes not, as in the case of Isabelle Dinoire)—but also on the basis of the life (such as it is) led afterward. Much as the trauma of the makeover process earns the participants in reality television the right to a new appearance, so too is an entirely new face earned by trauma. But the trauma lies not in getting the new face, nor even in losing the original face. It lies in the experience of life without one and overcoming its challenges, being ennobled by suffering and realizing its value. And only then being given a chance to avoid it.

It's about time, again. A new way of keeping time. Unlike cosmetic surgery, the face transplant doesn't erase time. It changes it. The new face marks the realization of a new self, a—so the narrative goes—better one. In the case of Isabelle Dinoire, as we've seen, there was not enough time: to come to terms with facelessness, to change (for the better) because of it, to demonstrate her worthiness to transcend it. In this chapter, I've explored the television news reports of face-transplant recipients, considering their similarity to traditional makeover formats and the stakes for these parallels. I've paid attention to the gender presentations and how they impact the packaging of recipients and the presentation of the procedure, turning my attention to the larger institutional issues these questions reveal. We thought about the role of the doctor as makeover specialist and charismatic constant, contrasting his or her (usually his) form of celebrity and expertise with that of the patient. And underlying all these themes was the always present and very real trauma, pain, loss, and transformation, and how that last piece is meant to alleviate the effects of the others. It may or may not do so—we don't always know—but it doesn't matter. Not for the purposes of the makeover fairytale.

That's a powerful, even elegiac kind of thing to write. And I mean it, and I think that it is right, and I hope I've convinced you of that as well. But it isn't entirely a matter of packaging and presentation, or, at least, it isn't entirely the reality-makeover packaging and presentation. Because (even as that genre has borrowed heavily from others) operations are about science. And knowledge.[77] Once it became widely accepted, the face transplant had to be thought of in a teleological context of continual beneficial improvement. Not the realm of mad experimental scientists and evil doctors. Not, despite repeated tropes of sexless rebirth and unfathomable technology-aided opportunity, anymore. Makeover television tames the potentially subversive elements of radical bodily transformation, as Meredith Jones has cogently argued.[78] The techniques of medical documentary make clear who the good guys are. (Hint: it's the doctors.) In the end, while we are rooting for the patients, they come and go. It is the doctors (and the expertise they are promoting and the industries they are supporting) who make it all happen. Who begin to make it all right for the patients, who solve the problem of these patients, in the end. And it is they, even as the recipients may fade out of popular memory, attention, and interest, who remain.

The doctors save our heroes, but only after these heroes have begun to save themselves, starting the process of making themselves over into their best personal version, made possible and necessary by their injuries. In this reading, it doesn't matter if the face transplant is cosmetic or medically necessary or both or neither. What it is, unambiguously for the faceless individuals and the society that ignores or demonizes them, is a problem to be solved. Even more so in the case of those injured in service of their country, as the military funding of the surgery makes clear. Faceless people are a problem, but faceless vets are a crisis—to national identity, to military expenditure, to our own sense of safety and stability. It is a crisis that is solved, or solvable, with a new, conforming face to accompany the new, acceptable, conforming self. The military gets that: that's why they've invested so heavily in face-transplant research, both experimental and clinical. That's why they continue to fund the surgeries, almost all of which are on civilians. And that is why, in the end, it doesn't matter if the recipients offer a radical way of reconceptualizing humanity. Because there is nothing radical about making injured vets fit to rejoin society, or, indeed, making anyone fit to rejoin society on society's terms. As we know from the makeover genre, which makes scars invisible, hybridity seamless, monstrosity unimagined, and abnormality—as defined by anything other than conventional beauty and standard gender presentation—eradicable. And eradicated.

But it could have gone another way. The resistance of the face transplant to categorization didn't need to be a problem. It could have been—it can be—an opportunity. A chance to embrace ambiguity and hybridity as a way to envision a radical form of personhood initiated by but not limited to those who have intervened in the stability of their faces and their relationship to character. It can be a way to reimagine the body and its stakes. A way to acknowledge and even underscore the importance of the face without being limited by it. A way to use the flexibility of the body to propose a third way to think about faces and bodies that is rooted neither in risk nor adherence to social norms. A way to mark trauma rather than disappearing it into the everyday. A new form of ethical interaction and relationality. What would that look like?

Let's imagine.

CHAPTER 6

Conclusion: Face Transplants and the Ethics of the Other

INTRODUCTION

This is it. The big payoff. The ta-da moment. (The big reveal?) My (humble) offering of what a radical ethics of nonindexical facehood would look like. A way to make real the possibilities—offered to (and rejected by) us—of the face transplant. And, actually, a way to make real the possibilities offered of all other interventions to the face, to the self. What is at stake here is not just changing how our faces appear and what that might mean. It's about prosthetics (do they make us different and worse? are we better than well with them?[1] are we still ourselves?), wearable technologies (does it matter if we can take them on and off?), online avatars (also wearable technologies? prosthetics? alternative identities?), and entirely new ways of thinking about the meaning of the self and the meaning of the self in relation to others, which aren't rooted strictly and entirely in the bodily.[2] But I don't want to do away with bodies or appearances altogether. As I stated way back at the very beginning, faces matter. A lot. And that's not necessarily a problem, and it's certainly not up for negotiation. I don't seek to undermine the kinds of identities that develop from, or contribute to, the way we look. Lived experience and the materiality of the body matter, even or especially as the body changes. So does history. The ability to self-consciously construct our visual identities as a reflection of our characters, and our inability to control how parts of our characters have developed from our appearances, are far from meaningless. To be more concrete: I don't want to slip into a naïve kind of a color-blind, gender-blind, age-blind mode of encounter that isn't, in the end, blind at all but rather ignores distinctions that are powerful and ramify and have implications that need to be acknowledged, and, sometimes, celebrated.[3] No. Not that.

I also won't offer a synthetic conclusion that neatly ties together all my arguments about the status of the face and the ways that is revealed and challenged through face-transplant surgery. I can't. There isn't one.

What I do want to do in this concluding chapter is show you how face transplants and the culture in which they emerged, and the ways that they and the people who undergo them are represented, have changed the way I understand theories of faces and, indeed, faces themselves. In so doing, I hope to voice my own intervention into the debates around the meaning of identity and embodiment, bringing my training and perspective to bear on the various approaches to these questions. As I said at the outset, this is a personal book even as it is an academic one. The theories that I turn to in this chapter as a way to reimagine faces, voices, and people are central not just to the themes in this book but to how I understand myself as a thinker. And how I understand myself as a person: as a media-studies scholar who thinks about bodies online and in the world, who thinks about popular culture as well as theoretical writing; as a feminist who cares about difference and intersecting identities; as someone trained as a historian of science, medicine, and the body; as a Jewish woman with a heritage steeped in hermeneutical relationships to biblical texts; as someone striving to be an ethical person in the world. These are all the reasons I care about (am obsessed with) the face. And the people behind the faces.

When I envisioned this chapter, I thought it would be an examination of what a Lévinasian ethics of the face transplant would be. If so, I would have focused on Lévinas's claim that "a face is abstract" rather than being tied to particular physical characteristics, which helps frame what it might mean to say that "the best way of encountering the Other is to not even notice the color of his eyes!"[4] Lévinas's eradication of the image of the face, while emphasizing the centrality of faceness, holds promise for a radical ethics of relationality that recognizes the changeability of the face and its constituent parts. The face is an object whose encounter leads beyond its objectness; it is unlike anything else.

To remind you, French philosopher Emmanuel Lévinas frames his ethical system on the obligations that emerge from the moment of encountering the face of the other. Lévinas postulates that (when) the other cannot be controlled, the response may be to try to subdue through murder. But instead we follow the injunction "do not kill," and from this moment we have the origin of our ethical obligations to the other. It isn't obvious, the ethical response. It is, instead, the mastering of impulses through the power of the other, which is the power to stop murder. Violence is rejected. And the origin isn't about knowledge, not yet. Any comprehension of the other through knowledge is totalizing, which is to say, anytime one takes an idea about the other to be the actual other, the ability to have contact with that person is denied. The autonomy of the other is denied. Ethics emerges from the connection with the real person, not the idea of that person. Only the phenomenological experience of the face provides nontotalizing ground. Provides responsivity. Provides transcendence. In this moment of encounter, this moment that is the origin of ethics toward the second other who is one and the third other who is all (of humanity), one also becomes oneself.[5] Prior to the encounter, there is no response. There is no ethical self, except in the encounter. There is no self except in the obligation to the other. The encounter with the other is generative of the self. The face-to-face encounter is a moment of intimacy, and obligation born of intimacy. And that which arises from it works in both directions: both parties are changed as a result of that moment. The viewer has a new obligation and becomes a new person as a result. And the viewed, the other, becomes a participant in the system. We need the face-to-face. For the other, and for ourselves.

But the Lévinasian face exists beyond any given face, beyond any specific face. For Lévinas, the face is the surface of touch and exposure. Face is not façade: it can be changed and changed again to save face or even gain face. The Lévinasian face overrides just that specificity and contextuality that I want to insist upon in the story of its change(s). I want, here, to think the ethical call of Lévinas with the kind of (Deleuzian) subjectivity that allows for individuality through constant change, that sees being as always becoming, that asks not what is the body (face) but what can be done with it or how it comes to be.[6] Specifically, I want to know here what an ethics of the ever-changing face might be. Which requires, as it must, a path through my own very personal matrix of ways of encountering the other.

I stand on the shoulders of giants. Many scholars and theorists have explored how faces work on the individual level, how the rhetorics of the face work on a broader scale, and what the posthuman face and body might be.[7] I want something concrete. I want to solve racism and sexism and allow for maximal self-definition with minimal self-limitation within the parameters of a new ethics of nonvisual encounter with the other. With all others. But I don't think I'm likely to do it here (or, let's be honest, at all). And, at the same time, I can't give up on transcendence.

I won't leave out the possibility or promise of transcendence. But, in the course of learning about face transplants and face-transplant recipients, my commitments have changed. My ideas about faces have changed. The way that I understand theories about the face—the theories that matter most to me—has changed. I thought the big payoff would be about theory, a new ethics of interrelationality and a new way to understand the face. And I still think we need that, and I think it is emerging, as these things do. But that's not the concluding narrative of this book, not really. (Though that would be synthetic. And nice. And maybe braver than I am.) Instead, it's about people. It's about the way we think about people, the way we represent them, the space we give them to speak, and the voices that we hear. It's about trying to find a way to fight the codedness of the face (any face, be it manipulated, transplanted, or aging and changing), a fantasy imagined through endless and ongoing rhizomic experimentation with meaning.[8] And that's a nice fantasy, but we can't persist in that endlessly explorative space, not really. So it's about offering livable possibilities by recognizing that wearers of faces and facelessness and their observers (which is to say, everyone) must always struggle with the reality that the face or its absence is *always there to be seen* and always opens us to the vulnerability of being seen.[9]

So. The face transplant can be thought as the most direct attempt to restore and stabilize signification alongside the attempt to recover the undefinable resonance of what is lost with the face itself. This book has been an attempt to trace the contours of that loss and how it is experienced and understood; here I seek to imagine the possibilities of living within the paradox of a face that is not overcoded but is, as a function of the experience of embodiment and the experience of history, still necessarily a part of the system of signification. In its final form, this is a story about real people with real stories and real cultural imaginaries, which means (as always), in the end it's about ourselves.

And it's about the other.

We're back to Lévinas again.[10] And Deleuze and Guattari, and Habermas, and Butler, and a whole lot of other people who thought about the face and the space for interaction with and being obligated toward and having knowledge of the other.[11] They disagree, these thinkers, about how we encounter the other, what role the face plays, and where the possibility for relationality lies. They disagree about how important the face is, and they even disagree about what kind of object the face is. But they also agree about some things, sometimes. Lévinas and Habermas both see great potential for ethical exchange in the communicative power of speech. Even Deleuze and Guattari, whose concept of faciality and the assemblage is a direct counter to the Lévinasian notion of face, converge with Lévinas around transcending or thinking beyond the visual in understanding what matters about the face of a person. For them, the face is the most iconic form of signification, with all its overcoded rigidity. The face, the lived understanding of the face, is precisely this constant effort to code the uncodeable that is the other, the intersubjective and ever-changing other. A coding that rests on the face as revelatory of something that cannot be reduced or thought, even as it must be encountered.

At the heart of all of these approaches to the face is the question of communication in the moment of encounter, and the possibilities for that moment and those that come after. Because encounter *is* communication. Which comes back, again, to the face.

The face-to-face.

This is also a Lévinasian formulation. And I'm not about to give up on Lévinas, or to abandon theory, not at all. But this book has taught me that changing conceptions of the face in the face-to-face is not enough. We must, somehow, parse the paradox of the totalizing power of the face, in its infinite weakness, while acknowledging the infinite possibilities of the self and the other. Challenging the index is not enough. We have to be attentive to voices where we can hear them—in their own stories and in the way the stories are told, represented, and understood. We have to be attentive to history and how it is inscribed onto bodies and how it is read into bodies. And we have to be attentive to how voices and history interact. Bodies and faces still matter, and the ways that we understand them still matter, but so does the way that people understand themselves as bodies and faces, and as the repository of their experiences and as the result of their pasts and the pasts that came before. And what I've seen is that people long to understand themselves, and to be understood, as always in the process of change. And some, as I show in this chapter, have made their commitments to process explicit and deeply visceral. They have expressed their voices through their manipulations of the body in order to challenge the primacy of the body itself, and the supposed indexicality of the face, and the limitations imposed through this indexicality. And to expand our notion of what the face is, and how many faces we have.

But still it remains: we (yet) don't trust the multifaced. The two-faced. Two-faced means villain (not just for Batman fans).[12] We want to know who we are talking to. But maybe we shouldn't rely on the face as the way to begin knowing, and maybe we shouldn't think about those opting out of the system (by manipulation of the face) or being forced out (by facelessness) as cheating the matrix of communicability. Maybe, instead, they are subverting its tyranny, forcing us to direct our attention elsewhere. Forcing *us* to direct *our* attention: the system of facial judgment is always a relational one. It's always one dependent on the public manifestation of the self, the viewing of the self by the outside world, and the power that emerges in that space. This whole book is about the face, but it is also in a way about the making of the face—the makers, the ones being made, the ones for whom it is being made, and the space in between. In this chapter we'll think explicitly about process—specifically, we'll think about the face as always in process. We'll think about what it means to always be becoming.

While my focus is on the face and its special status, I will touch here on the body and the bodily more than I have before, thinking about faces in particular and corporeality and its limits more generally. I'll think about possibilities, but also limitations. Not for here the "postmodern imagination of human freedom from bodily determination" that Susan Bordo has critiqued so cogently.[13] She shows how the realizations of these fantasies erase difference and offer an often impossible and certainly undesirable standard of the feminine ideal. Nor do I suggest a utopian futurism that elides differences in lived experience or bodily desire. Even better-than-perfect bodies and faces are problematic sometimes: prosthetics may be more trouble than they are worth.[14] People remove breast implants. Bodies determine, and faces—with their special status around identity and communication—even more so. But maybe we can change how. Or maybe we can at least identify the power relationships enrolled by facial determination, the publicity of faces, and the stability assigned to their indexicality, and maybe then we can begin to destabilize them.

IRL 1

But there are a lot of different kinds of face-bodies. Do digital bodies matter in the same way as those IRL (in real life)? Many have dealt with this question, and research into digital bodies, experiences, and identities is rich and ongoing. These debates, of late, have shifted from the early questions of identity laid out by Sherry Turkle and Howard Rheingold into explorations of the digital experience as one of play and performance.[15] That shift happened partly because the context of the online world has changed with the explosion of the commercial Internet, as Lisa Nakamura discusses in her analysis of tribal marketing and the ways that people understand themselves through commercially structured and motivated interfaces.[16] Self-expression online, Yoel Roth argues, is now explicitly undergirded by structures designed to make our digital selves as lucrative as possible.[17] Which means that the vexing of the index between face and character in digital life doesn't really matter, at least not for the bottom line. Or, better said, the mapping of a digital self onto an IRL self isn't really the point as long as the digital self is monetizable.

The real-life encounter with a person known only online happens, particularly in the arena of geosocial networking applications and digital worlds and games. The online environment is another space where the index between specific face and particular person has been disrupted, in ways that almost always create both problems and possibilities.[18] Yes, as Sandy Stone tracks in a now classic essay, disabled older women online can actually be able-bodied men, but they can also be differently abled women with new capacities in the digital space.[19] (And, as we know, that's far from the worst or most chilling case of online identity fraud.)[20] Yes, the affordances of the digital allow unprecedented forms of sociality and access for large communities of people.[21] (Self-selected, it must be said. For the formerly faceless people that I examine, a life lived exclusively online would hardly be enough.) And, of course, *yes*, we turn to these very commercial infrastructures that profit off of our self-expression to provide some kind of suture to the crisis of indexical rupture. That's why one has to be a "real person" on Facebook (though the parameters of that, too, are changing, as the recent decision to allow drag queens to have personal Facebook pages shows) and why so many social networking apps are indexed through Facebook in a process known as bootstrapping.[22] Even online identity is policed; there may now be sixty permutations in the gender field on Facebook, but dating apps, including the lesbian-targeted Her, use Facebook to ensure that one is "really" a woman. One may have entered something else in the Facebook gender box, but its apps will nevertheless scrape the Facebook backend to decide in which of the two gender binaries one belongs.[23] Self-representation online may invite performance, but monetizable identity is still very much in play. And being verified.

We may imagine that there is a whole (digital) world out there (in there) that doesn't care at all who you are outside of it. But that world does care about corporeal you in so far as it can use your data. It will market to your affective online identity whether or not it matches up with your personal performance IRL. Multiple identities, in that world, are even more valuable: they provide more options. In that world, it isn't that nonindexicality isn't a problem. It's that the index is infinite; you actually can be many people all at once. You can have a burner account (or two or three) on Kinja and be a user on 4chan and an avatar (or two or three) on World of Warcraft, and they can all be the same or different genders, races, and ages, or they may not be human at all, which in no way undermines your actual humanity. And you can log off and be someone else entirely. And it only matters if you say so. Or, really, if someone else does.

We have something to learn from the digital anyway. Maybe it matters because it is a space in which multiplicity, complicated as it may be, contested as it may be, can be. The index is vexed, and we carry on. We change the conversation from one about identity to one about performance, which is a recent scholarly move but hearkens back to a very old conversation. Sennett already knew that we performed ourselves.[24] So did Goffman.[25] But there is something particularly . . . strategic about the Goffman version that Lévinas (as we'll see) undercuts through his insistence on openness to the other. Something, as Smart argues, that is calculating. Amoral.[26] In Sennett's and Goffman's formulations about the public performance of our identities, about the presentation of our particular and changing faces, we couldn't—quite—walk away from these versions of ourselves. We couldn't leave our identities behind. We still can't in real life, not really, not without leaving a whole lot of other things behind as well, like our relationships and our families and our lives. We can briefly play roles with the fully self-conscious adoption of those characters, but that also isn't the same thing as inhabiting multiple online identities. Mostly because there is a limit to how long we can play roles IRL. We can role-play ourselves into other people, but those other people *are* role plays, separate from our real selves. Online, everyone is always already assumed to be inhabiting a role. Online, playing *is* real life, with real financial implications.

The stakes are different for being faceless online.[27]

Whatever that means.

Amit Pinchevski has an idea: he thinks that chat rooms allowed subjects to turn away from responsivity to the other, whomever the other online may be.[28] There can still be encounter online. There can still be obligation. But without the face-to-face, without the phenomenological experience of that encounter, there is distance. Roger Silverstone's "proper distance," maybe, a distance that enrolls an ethics of care by imagining the other in her own terms; a distance which Introna would say contributes to Lévinasian justice by making all equal through the proximity of the screen.[29] But the screen also creates the distance that enables us to ignore the enjoinder to respond to the other. To make the other into something else. Something that can be ignored. Except John Durham Peters has taught us that we—not the machines and not the animals—are the other, and that we should value one-way communication, from person to screen, from person to the world.[30] And he would ask: what about touch?[31] What about the ability to touch and be touched and feel and be felt? Dallas Weins couldn't feel his daughter's kisses until he got his new face. That's one of the reasons he did it, Weins tells us.[32] Does the attention to lost faces and their recovery turn us back to embodied communication, back to the body as the way of evaluating the success of communication? Back to the physicality of keeping in touch, and the importance of that physicality? Or, what is communicatively denied by the inability to touch and feel, either because of a mediating machine or because the sense itself—experienced bodily and trained socially—is gone? Communication is still possible, but—so Peters would say—something is lost. Dallas Weins's account of his life of nonhaptic loneliness, of the distance that could not be bridged by touch, shows that to be so. And with the face transplant, with the renewed sense of touch, with the ability to feel the kiss of a child, the embrace of a lover, something—maybe everything—is gained.

If you are faceless and no one can see you, does it matter? Or, to put it another way: are faces ever anything but public?[33]

This is a question about how power and influence are embodied and expressed, and it's also a question of relationality. As we'll see, if the face is covered, then not only is the other denied, but so is the subject. Covered faces elide the system of ethical response and deny humanity. But a faceless face is not a covered face, nor is it an unviewable one. At least, it ought not to be.

I'm not going to get into the questions of affective life online; I'll leave the meaning of the rape in cyberspace and the implications of attacks on digital bodies, virtual bullying, and app-romance to others who recognize the possibilities and the challenges.[34] But I will say this: we have begun to imagine and even to experience a way of being in the world that knows about faces and bodies and cares about them but cares about other things, too. It's not perfect, not even close, but it's a start. So let's start there.

What can we learn about changing faces from a space where the face itself is infinitely changeable? (But, importantly, not interchangeable. Specificity still matters in any given moment.) An avatar is identifiable, and it belongs to someone, and that someone is sort of the user behind it, and sort of the platforms who own and license it. This, at least: just as you can have a face with no person (or a face with many people), you can have a person with no face, or, more exactly, no face as we traditionally understand it. There are people behind the sometimes visual, sometimes textual, sometimes both, avatar. And there are people behind the faces-that-are-denied, those who have lost their faces. They aren't people only when they get their transplants. They aren't necessarily new people then, either. And there are people with many online faces, all at the same time. These different takes on the face cast the meaning of the face and its indexicality into doubt. But we already knew that appearance doesn't correlate directly with and inherently with behavior and character. We've left behind the legacies of the physiognomic practices that linked the internal with the external in concrete and deterministic ways without the help of the Internet.[35] Haven't we?

No, of course not.

Straight talk: we judge the hell out of people for how they look, for how they present themselves, for the color of their skin, the performance of their gender, their adherence to social norms, their (dis)ability, and their inhabiting of age categories. We think how they look not just says something about who they are but is who they are. And we're not wrong, at least in so far as some of that presentation is a reflection of a personal (and changeable) choice informed by histories both personal and general. And we're not wrong, at least in so far as that identity is formed not just through personal choices but through interactions with others. And we're totally and completely wrong (wrong—not just in an empirical sense but also an ethical one) in that people change. People are always changing, on the inside and the outside. And, as we'll see in my discussion of the performance of plastic surgery, we judge change too.

The index has always been vexed. Face-transplant surgeries didn't invent the possibility of radical hybridity or the instability of identity. But they did put them, in this most concrete of manifestations, on television.

So let's take that vexing seriously. Let's posit that appearance matters in the formation and expression of identity and experience but need not be the sole determining factor of it. *Is* not the sole determining factor of it. And let's go further: let's say that the extent to which appearance, and specifically faces, correlate with who we are should always be treated as an open question. One that we can answer only through further engagement with those whom we wish to know. Those whom we should not have already judged. Which means, in the first instance, that we have an obligation, brought into being through the very act of looking at someone else, to respond to the other.

I will propose what the right question might be (indeed I have, in many ways, throughout this book), but first I want to discuss what happens not to the faceless but to the faced in the moment of encounter. I want to think about the reflexive experience in the Lévinasian moment. Put simply, in what way do I become "I" when I respond to the other? It is in that question that we might begin to understand what the true potential of the face transplant or nontransplant might be for reframing how we understand identity and its relationship to the body. We might begin to answer Donna Haraway's famous question, "why should our bodies end at the skin?"[36] Simply, according to Lévinas, they don't.

HOW MOSES DID IT

And since I'm being modest, I'll start with a modest moment of encounter: the (failed) face-to-face between Moses and God. (I'm joking here, which is how I'm being serious: modesty is a form of effacement. I withdraw my face, and my voice. I turn it over to another, more enduring and more powerful text.) It's the story of an incomplete face-to-face (f2f); Moses, lonely, desperate, and yearning for a moment of intimacy with the divine to whom he'd dedicated his life (sometimes at great personal cost and often with great emotional and physical pain) begs God for a glimpse of God's face. The chance for a real conversation.[37] A moment of meaningful and almost equal encounter. Moses wants not to know God—he already does—but to experience the Lévinasian moment of preknowledge. Moses—he who was closer to the divine than anyone in human history, he who was selected and trained to redeem an entire people, he who always only felt inadequate and overwhelmed—wished, in this fantasy of encounter, for transcendence.

He didn't get it.

He got intimacy. But he already had intimacy: the text states in Exodus 33:11 that "God spoke to Moses face to face, as a man would speak to his friend." But just a few short sentences later, after Moses pleads (indeed, the text says, begs), "I pray you, show me your glory!" God responds that "I will allow all my goodness to pass before you," but that Moses cannot see it. God says to Moses that God will "cover you with my hand until I have passed by," because "no one can see my face and live." God will "take my hand away" after the divine face has passed by, leaving Moses to see only "my back."[38] It's a powerful and embodied encounter, but it is not the face-to-face. And not even totally embodied; eyes can be covered without touch, that most fundamental way to speak without words.[39] The Babylonian Talmud teaches in Brachot 7b that, even here, touch is denied; Moses, in yet another moment of distancing, was covered by the knot of the phylacteries. The material representation of the divine, perhaps, but one that is available to all (men). This passage teaches us that, as intimate as God and Moses were, as much as they spoke to one another as friends, it never was true transcendence. Moses would know God, but he would never, in Lévinasian or maybe Buberian terms, preknow him.[40] Their relationship was never one of ethics. It never could be. For Lévinas, God's passing leaves a trace, the trace that is found in the faces of humanity who were created in the image of God.[41] To approach God is to encounter humanity. But that is not the encounter that Moses wanted. Nor, indeed, the one that was asked of him. Moses wanted touch, and he got the Law. He wanted intimacy, but he got justice.

There was always going to be an imbalance in this relationship, between humanity and divinity generally, and Moses and the Israelite God specifically. Hegel would put it in terms of the master-slave relationship, one in which both parties were complicit and in agreement and in which each failed to achieve total self-consciousness.[42] To Hegel, there is a narrative of power here, and consensus around the power disparity and attempts to forge a relationship going forward. Lévinas would disagree: there is no dialectic in the encounter with the other. It is not a new combined entity that emerges but a distinct third, namely a responsibility not just toward the immanent individual but toward humanity.[43] For Lévinas, power is in potentiality; in every act is a loss of power. God doesn't need recognition, and Moses couldn't grant it. But Moses wanted it. Moses wanted the Lévinasian encounter with the divine other to give him the opportunity to respond, and in so doing to fully become himself. It is the opportunity to realize the full meaning of the "hineini" moment, when Moses responded to the call from the burning bush with the statement "here I am."[44] But without the face-to-face encounter with the caller, without the ability to see the face within the flames, the statement, again as Buber teaches, is untrue. Without the true encounter with the face of the other, there is no fully realized I.

It is in the face of the other, the response to the face of the other, that we finally become ourselves. It is in the faces of the others that we are revealed; without them, we cannot see our faces. We can never fully see our own face in the face-to-face. I can see your face and you can see mine, but I can't see mine and you can't see yours, unlike almost every other part of the body. Our faces are obscured from ourselves, even as they are on display for the world. There is possibility in this blindness, the chance to embrace the other as a way to be and a way to see. There is danger in this blindness, the danger of valorizing the visibility of a particular kind of face of another as a way to compensate for what we cannot see in ourselves.

But face is flexible, abstract, and not about a collection of features. Not in the Lévinasian system, and not in the world of the disfigured and the dissatisfied. Not in the world of the faceless. So it shouldn't really matter that Moses saw flames instead of a face, a back instead of a face (because faces need not be strictly facial). Except that the face was deliberately obscured. Moses was denied the faciality of the other, and he failed to recognize the face that was present even in absence. Moses failed to see the face of the other. Moses failed to truly respond. He failed, in the context of the response to God, to become truly himself. There is no ethic here. (Nor, some commentators claim, is there an ethic in the relationship of the Israelites to God. Though they accepted the commandments at the foot of Sinai, they did so with the mountain held upside-down above their heads. The agreement was born of threat and self-protection, of the knowledge of sure destruction in the case of refusal. There is still justice, the origin of justice, in the commandments. And justice leads to the face. But no transcendence here, not yet.)[45]

So what then of the obscured divine face, the impossible-to-imagine face? Can we make a meaningful distinction between the unknown face and the unface? Is the encounter with the faceless the encounter with the divine, that which never calls a totalizing knowledge into being in its imaginary, because none can exist?

Well, no. Even the faceless have faces that we can see. They are just different. And note: I'm not criticizing the decision to veil a disfigured face, nor the deadpan response. The strategies of challenging the system highlight the flaws in the system itself.

Even as the face itself is a universal abstraction (or maybe even, according to Deleuze and Guattari, nothing specific or transcendently unique at all, an overcoded imposition that is meaningful only in that its degrees of deviance from the originary white male face is the standard by which people are judged), it engenders a relationship.[46] Maybe *because* the face is an abstraction, it engenders a relationship and an ethic. But the Lévinasian face is two-faced; it can never fully efface its origin and mark. It is corporeal and specific. Singular. And also each face invites infinity, responsibility to the infinity of all humanity. We approach the other beyond the face, but the face itself stops us short, always reminding us of the specificity within the infinite. Surface and face: immanence (to the specific face) and transcendence (to all). It doesn't matter whose face it is. Or even how it came to be a face. And it does matter, the face itself reminds us that it matters, reminds us of the specificity within in the infinite. And it does matter, in the Deleuzian sense, because history matters, and the history of the face is the history of racism. I don't seek to evacuate that past but to complicate it by imagining a face that is not forever being compared to an imagined ideal face, by imagining a wider definition of the category of face. The faceless, the facially manipulated, and the face transplant are variations for Lévinas. They are all still faces (and maybe and maybe not surfaces). They all still demand a response. They all still invite transcendence.

The human (because they are human and not divine) faceless still demand a response. Let's posit a world in which that response is not murder.

It doesn't have to be silence either, and maybe it shouldn't be, as Rosemarie Garland-Thomson has argued in her work on staring.[47] Acknowledgment of difference acknowledges that we are indeed all different. And then, there is no standard. There is no normal. It's a nice argument, and a stirring one, that speaks to Bordo's critique of the fantasy of infinite human possibility facilitated through technological interfaces with the body.[48] That would work if, as Bordo highlights in her analysis of the postsurgical body and presentation of Madonna, the possibilities didn't end up adding up to attempts to achieve exactly the same, generally impossible, highly gendered ideal.

It doesn't mean that Donna Haraway was wrong when she said it would be better to be a cyborg than a goddess.[49] What is missing is the caveat "one day." We're not there yet. According to Bordo, Shildrick, and Garland-Thomson, today most cyborgs are still doing their best to be (gender-compliant, physically attractive) goddesses.[50] What is radical about an intervention that, instead of facilitating change, attempts to reinstantiate an even more sophisticated version of the same old thing? And I'm not—here—critiquing the same old thing; let people look how they want, if that is indeed how they want to look. As I explored in the second chapter of this book, it's too easy to critique people for trying to adhere to a certain ideal. Imani Perry has shown us why women attempting to look less black through skin lightening might be a good strategy in the particular, though with general (negative) consequences.[51] Kathy Davis has warned us about silencing women's voices and questioning their needs as they articulate them.[52] It's complicated. Equally complicated is that not everyone has access to the ideal (either naturally or through interventions), and some/most are penalized for it. A radical intervention into the stability of appearance is only radical if people think it is. A lot of work has gone into neutralizing the face transplant and making it deeply normative, like most other forms of surgical and prosthetic interventions into the body.

But not all. There are those who do it a different way. The explicitly radical performances of the surgically and prosthetically manipulated self, and the ones whose manifestations of the self offer a different kind of normal. Or (maybe), in Haraway's terms, a different kind of superhero.

IRL 2

Superheroes and radicals are both equally fun, so there's no clear way to decide who goes first. But there are probably fewer superheroes than radicals, so let's start there.

In the British spy spoof film *Kingsman: The Secret Service* (2015), the femme fatale villain, Gazelle, uses her two prosthetic leg-blades to devastating effect.[53] Unlike her opponents, she never needs to reach for a weapon; she has two always at (er) hand. But her leg-blades are neither what make her evil nor what make her intriguing. She's a character in her own right in the grand tradition of female secretaries with minds and powers of their own (see Potts, Pepper) on whom, in the end, the success (or failure) of the venture partially rests.[54] She's a sidekick, a bodyguard (in a nice bit of gender play), a beautiful woman, and an evil superhero with the power to best the good guys. With blades for legs. With blades for weapons. And she's notable because she's just another one of the superheroes in the cast, each of whom is devoid of supernatural powers but has all kinds of neat technologies and impressive skills. Gazelle is also an artist, whose carefully choreographed and beautifully sequenced moves and movements speak to the actress Sofia Boutella's training as an elite international dancer. As the film's director and producer Matthew Vaughn said in an interview with the *New York Times*, "we've created an iconic, cool disabled person."[55]

Indeed they did create her. In another grand tradition of adding difference artificially, the blades were added digitally. Sofia Boutella has two traditional legs.[56] Suddenly the character, as an embrace of the multiplicities of normal in the world of the superlative, becomes a whole lot less interesting. It's still cool, and maybe a little iconic, that there is a villain with blades and mad skills with them, but it doesn't really change the conversation. Of course movies aren't really life, and no one expects the characters to possess their onscreen skills and powers offscreen, but maybe, in the case of spy movies populated with real people who can do extraordinary things, we might expect them to possess their same bodies. And most of them do. Just not the one with the blades.

There are of course numerous examples of people with prostheses doing extraordinary (and very ordinary) things.[57] But what is so disappointing about the Gazelle/Sofia Boutella case is that it is a missed opportunity to make real the possibility that physical differences can be both disabilities and chances. It's not always true, and maybe even rarely, but here it could have been. Here we could have had Haraway's cyborg. Instead we have, again, Haraway's goddess, digitally altered.

IRL 3

Does the cyborg-as-goddess exist? (We're all cyborgs, of course; we wear contact lenses, take medications, attach devices, have artificial parts and bits.)

We can ground the question of vexed index in those who have made real the potential of the infinitely changeable exterior self in an explicitly public way. Publicity can be both radical, as in the performative surgeries of artists like Orlan, Lolo Ferrari, and Genesis Breyer P-Orridge, and (as Bordo argues) deeply normative (or groping toward normative), even as it queers categories of bodily boundaries around age and desire (in the case of Joan Rivers) and gender, race, and sexuality (in the case of Michael Jackson).[58] Publicity itself need not be radical; as age-erasing procedures become the norm in certain contexts (say, that of the *Real Housewives* franchise), the radical piece would be foregoing them altogether rather than refusing to put them on display (as Jones claims).[59] What is radical is the performance. The play. The instantiation of the changeable self not only in behavior but in body. Which makes real the abstraction of the many faces that Goffman says we are forever wearing and switching, and which not only vexes the index between face/body and character, but also fundamentally changes the meaning of the former. As Warwick Mules has argued, the singular specificity of the face is that which allows it to become something else.[60] The singular specificity of the face in any given instantiation is that which makes real the possibility for change.

A lot has been written about Orlan and Genesis P-Orridge, both of whom deliberately use(d) their bodies and faces as canvases for their artistic interventions.[61] Analyses include questions of their relationship to normative body types, whether they are triumphing cosmetic surgery, what kind of gaze they enroll, if their practices are engaging in capitalistic economies of surgical self-improvement, the role of pain in performance, and if bodies ought to be artistic sites of intervention at all. Most of the work focuses on French surgical performance artist Orlan, who uses her body as a canvas for multiple creative plastic surgeries, including for example the 1990s series *The Reincarnation of Saint-Orlan*. This project recrafted Orlan's body and face with elements of famous female subjects in classic works of art. She has been vocal about her embrace of surgical practice to subvert the dominant male gaze and reclaim agency around her body through conscious manipulation of traditional surgical norms and modes.[62] She documents her procedures and remains conscious throughout, disavowing pain, corporeal boundedness, and limitation, remaining always in a state of becoming and healing.[63] She and her surgeries make real a multiplicity of selves, facilitating a posthuman position wherein all these selves interact. It's painful (Orlan denies pain as part of her denial of the body as anything but a representational vehicle), but it's possible.[64] Orlan's project is a clearly and thoughtfully stated one that has been chronicled both by her and by those who follow and study her work. She is a member of the artistic elite, if a radical and controversial one. She firmly inhabits the world of performance and play, though her goals are quite serious.

Genesis P-Orridge is also an artist of the body who challenges corporeal limitations and embraces surgery as a means of surpassing and subverting them. Starting in 1996, the duo of P-Orridge and (the now deceased) Lady Jaye began their "pandrogeny" (positive androgyny) project, in which they underwent numerous surgical and pharmaceutical techniques including breast implants and hormone replacement in order to resemble and ultimately become one another, a single pandrogynous being. According to P-Orridge, rather than being "a male trapped in a female body" or "a female trapped in a male body," "I feel trapped in *a* body." Specifically, a sexually binary body; for P-Orridge, the "hermaphroditism" for which the pair strive "is emblematic of free existence. The notion of hermaphroditism has evolved to undermine the long-lasting dictatorship of the so-called normal."[65] P-Orridge is different than Orlan, however, in that P-Orridge uses surgery to overcome bodily limitation and (ideally) achieve a final and better state. For P-Orridge, surgery is the means to an end. For Orlan, the surgery itself is the point. And there is no end.

Orlan's surgery, Orlan's body project, *Orlan*, is never complete. It has no beginning or end, no before or after.[66] The surgery and the modification of the self are always ongoing, not because of the limitations of the technology to achieve the desired final state but because the desired state is one of never-complete, of infinite change. Of, in the Lévinasian sense, infinity. Orlan seeks to represent both immanence and infinity in one being, using her body to challenge the category of body and, in so doing, using one to stand in for all.

Like Orlan's project of continual modification, the face transplant is also never complete. The transplanted face is always healing, yes, but it is also always in an indeterminate state of being, always about-to-be-rejected and always being managed and subdued. As much as recipients (most, not all: remember Isabelle Dinoire) accept and assimilate their new faces, they are continually locked in a battle to assert dominance over them. It's the internalization and endless iteration of the Lévinasian originary moment—do not kill. Even when all your body wants to do is destroy (parts of itself). The destruction is controlled with antirejection medication, but the recipients live endlessly in the question. They too are never complete. They too are a multiplicity of selves (if unembraced), always encountering the other (paradoxically, there is no self in Lévinas outside of that which responds to the other, but here the self is other), always and forever becoming.

Can ethics be expressed in biology? And is this the root of why we are so discomfited by the face-transplant surgery?

The stakes for being in the state of becoming are precisely those that Orlan lays bare and those that P-Orridge, in the quest for a kind of embodied transcendence, denies. We (most of us) deny them too. We treat our physical selves—our always changing, sagging, weight-fluctuating, skin-tanning, hair-growing, shedding, dirtycleansmellygrossgroomedwrinkledgrowingshrinking selves as complete. Our faces are our faces and we are our faces, and, if they change (a little) and we change (a little), the changes probably somehow magically sync up. Because the ways that the face is constantly changing mean that even as the face is not at any given moment a perfect representation of the self, there is always the possibility that it could change to become so. Because the face is an index. Of something. Orlan's face is an index of something too, as is the face of the FAT recipients. It's not a stable index. (No bodily index is stable.) The state of becoming, the permanent state of becoming, is a deeply unsettling one; it is, as Orlan makes clear, gross. It's messy. The state of becoming mirrors the messiness and multiplicity of all acts of transformation.

The permanent state of becoming denies the stable state of being. Even as it requires it.

Orlan's critics, including the Australian body artist Stelarc, have argued that her dismissal of lived experience, the ways that people understand themselves with, through, and mediated by their bodies, exposes her adherence to a traditional Cartesian dualism. Orlan seeks to discard the body; Stelarc uses technology to redefine it.[67] And that's the conundrum, in a nutshell: how do we honor our relationship to the body while not being limited or defined by it? Or, better put: how do we encounter the face of the other in such a way as to honor the face and look beyond it at the same time?

Lévinas would say: we are always already doing that. Immanence and infinity. You the other-as-you, and you the other-as-everyone.

But we are not always doing that. We almost never are.

IRL 4

That's why Michael Jackson and Joan Rivers (for example) were also always in a state of becoming, a state of change that made recognition the question itself. Because they could never reach their ideals. But they sure were trying.

Even as both Jackson and Rivers played (quite, quite playfully) with the stability of the body, and even as they challenged the boundaries around gender presentation, sex, aging, and desire, they were still very much operating under a normative framework and a unitary sense of the self.[68] They wished neither (in Stelarc's terms) to discard the body nor to embrace its posthuman modification. They wanted, in the classic makeover formulation, in the language that has established bodily ideals and attempts to reach them, to be most themselves. To make their exteriors reflect their interiors. They wanted to change in order to remain or to physically realize who they were, be it reversal of age, gender, or race—they were not erasing boundaries but redrawing them. And while their interventions were, at times, quite radical (alongside other radical high-profile plastic-surgery players like Cindy Jackson the human Barbie and Jocelyn Wildenstein the tiger woman), their goals were very much not. They wanted to be themselves, and not anyone else. They wanted, even in their seemingly endless state of becoming, finally to become. Except, maybe the sheer fact of being oneself in a liminal state is itself radical. Maybe blurred boundaries as the final condition is that which we truly can't tolerate. And, again, the "we" matters; this is a public project, always.

Joan Rivers wanted to hold on to her sexual desirability and attractiveness. There is something defiant in her insistence that sexuality does not disappear with age, and something equally defiant in her fight to reflect and perform her sexuality over time. And she was quite serious in her project, even as she was able to laugh at it and herself quite explicitly and quite publicly, even allowing herself to be the subject of an episode of *Nip/Tuck*.[69] Her plastic surgery became its own topic, and she used her body and its manipulations to great effect in her comedy and the construction of her public persona. She frequently made jokes like "I wish I had a twin so I could know what I'd look like without plastic surgery" and "[t]he only way I can get a man to touch me at this age is plastic surgery."[70] But rather than framing her interventions as radical, for Rivers they were the concretization of the new normal. She believed that surgery would eventually become utterly mainstream, and she was only slightly ahead of the curve in embracing the tools of technology to eradicate the signs of age. The queering of social sexual norms is a byproduct of her pursuit, but also an indirect one; her goals were deeply normal, even as her methods were rather extreme.

Michael Jackson is also an extreme case. Kathy Davis has argued for the destabilizing nature of his iterative surgeries as challenging the boundaries around race, gender, and sexuality.[71] Jackson physically performed multiple intersecting identities while at the same time putting the limitations of the body itself into question, a point he emphasized in his transformations from human to werewolf to zombie in his iconic performance in the *Thriller* video.[72] The moments of morph in *Thriller* are a joke, a wink to the audience that invites them in to participate in his self-conscious performance (and critique) of traditional black masculinity and its campy disintegration throughout the course of the video. And Jackson issued the same invitation throughout the course of his life, as he surgically blurred the lines. He didn't talk about his project, at least not in the same way as Orlan, but he included us in his narrative of blurring boundaries. He told us in his songs that it doesn't matter if you are black or white, and he showed us in his "Black or White" video, with early use of morphing technology to show change as interchange.[73] It's a kind of a game for Jackson, but it's one that he doesn't really win. He never could win; as a black man of ambiguous sexuality, he was always already a freak.

Jackson's legacy is both as the King of Pop and as a makeover monster. His external transformations, commentators posited, must have mirrored deeply pathological internal ways of being.[74] The index, again. Even as Jackson can be read in consonance with Orlan (as indeed Meredith Jones does in *Skintight*) as another kind of bodily performance art, he is still very much measured against an imagined ideal that he was trying to attain.[75] Rather than see his transformations as an attempt to exist within the liminal spaces, to embody the also/and, Jackson's memory is very much tied into the either/or and, ultimately, the (failed) neither.[76] Of course, he failed to the outside world; questions of the body will always be mediated in the public sphere. Even if we accept that Jackson's final state is one of always becoming, even if we make the radical move, it is still more of the same, flipping the script that already exists. Jackson's body is still within the bodily framework. Bodies still matter. The stakes for appearance are still high.

Jackson is certainly an extreme case of surgical hybridity, and perhaps the very extremity of his procedures is that which made it a failure, highlighting the normal as that which Jackson failed to achieve.[77] Unlike Orlan, whose performance is designed to explicitly showcase the surgery itself, for Jackson (as read through his critics) the surgeries are still a means to an (ill-defined and unknown) end. We refuse to accept bodily performance art from this most iconic of performers, perhaps because he was a celebrity first and a bodily manipulator second. Perhaps because, unlike Orlan, he never told us that performance is what he was doing. Perhaps because, in the end, intent matters less than the final product, and the final product of Jackson's work reads less like a critique of the process and more like a failure of the process, a walking warning sign for the dangers of plastic surgery and hybridity and the proof text for what Jones has called "beautiful aliens."[78] And also because Kathy Davis was right: Michael Jackson's unsettling of the boundaries of the body and the boundaries of the bodily categories we impose is itself deeply unsettling.[79] We like our icons to stay where we place them, even if they have a little more leeway to play with their boxes than most others. We cannot accept the blurring of boundaries as itself the final product. Orlan's work is about the never-ending process; Jackson, perhaps, was arguing that the blurring itself could be a final state. In its way, this is more radical than Orlan, whose liminal position is one that is always in flux. Jackson's body suggested that liminality itself could be the end game, and he wasn't shy about it. He put his (monstrous, hybrid) body very much on display. Unlike Orlan, he didn't talk about it, so we didn't know quite what to do with it. And so we turn it into something else. We turn it into the kind of radical intervention that we know how to deal with and that we know how to either emulate or dismiss. We turn it into the (failed) attempt to become a god(dess) rather than the realization of the promise of the cyborg.

That's the kind of radical plastic surgery we are used to, the kind that goes to any lengths to achieve the impossible norm. We may laugh at it, sometimes, and we may desire it secretly or not so secretly; we may be entertained and repulsed by it (think of *The Swan* and *Extreme Makeover*), and we may critique it (remember the reaction to the new face of Renée Zellweger?), and we certainly, at times, expect it.[80] What we don't do, nearly as much, is imagine what it could be like. Most of us don't think of Michael Jackson in conversation with Orlan. Or with face-transplant recipients. Or with critical disability studies. Or with Lévinas.

Let's.

IRL 5

I want to hear the voices of difference and disability and disfigurement. I want to respect those who want to change, and those who very much don't. I don't want to tell anyone how to feel or how to be or who to be. I don't want anyone to have to earn their right to look a certain way or not look a certain way.

I also don't want better looking people to always get more nice things. (We define better looking according to a pretty stringent, pretty racist, pretty expensive, and pretty privileged ideal.)[81] I don't want people with facial disfigurements to feel like they have to look normal in order to function and be happy, but at the same time I don't want them to feel bad for wanting to change how they look. I don't want anyone to feel bad for that. We're always changing how we look, voluntarily and otherwise. We're always vexing that damn index.

Fuck the index. It shouldn't work. It only works because we say so. It only works because we treat people differently according to how they look, which is what makes the index true. We maybe knew this already, but we learn it more and better from the face-transplant recipients, who were people before their disfigurements, and people while these deformities were visible to the world, and people when they got new faces. Transplants made them easier for us to look at and for us to understand. Lévinas would say, but that's our fault, isn't it? We didn't then understand what the face is, or the role it plays in creating the space between. But what is left now, in the wake of face-unface-face, is to figure out what power remains for the face as a tool for becoming in the wake of the disruption of its magic. And the answer lies in how we handle the loss of signification in a very real sense. The answer lies in insisting upon the face-to-face-or-unface as a way to realize ourselves through our realization of the other.

That's the first step: divorcing the face from that which lies beneath. Getting rid of that particular index entirely. It's the hardest step in a way because we don't even know that we are adhering to the index (we are), and the easiest in another because, c'mon, none of us *really* believe in physiognomy, do we? (Ha. We believe in physiognomy. And not just "beady-eyed criminals." What is racial profiling if not modern physiognomy?)

Step 2: See the other. Also the easiest and the hardest. (After all, Lévinas said this, but he never told us how.) It's hard to look at the other, especially when the other is hard to see. But something happens when we look. We can no longer kill and destroy. Not easily. Ideally.

Step 3: Don't be an asshole. Really also should be easier than it is. Overcome the Lévinasian coincidence of the face and the possibility of violence. Instead allow the ethical encounter to emerge. Understand that the stakes for being present are very high indeed.

Step 4: Allow yourself to be seen. Think about that. It's quite hard. It entails a vulnerability and an openness. Also someone might want to kill you when looking at you. Do it anyway.

Step 5: Create the conditions of possibility for a space to emerge between you and the other. This is where we leave Lévinas. Lévinas got us here, got us to understand that the faceless still have faces, can still be seen. Should not be killed. Do enjoin an ethical response. Lévinas got us to understand that the face itself is an abstraction, a necessary one whose specificity matters only in so far as it designates one among many, but only for that. Lévinas taught us that looking at the other is about seeing beyond seeing. The features are irrelevant. Lévinas got us to the point where we can begin to know otherwise. Now—

Step 6: Focus on the space in between knowing and seeing. The face still matters. The body still matters. But for those living within them. Not for those looking at them. (This is really hella hard.)

Step 7: This step depends on what the ultimate goal is. But I think it is important to think about the kind of space that is formed in between, when an encounter allows for addressability. Habermas would say it is always and only about the space in between, the space of rational discourse. But discourse isn't always rational, and it is always framed by lived experience. I think it is enough that we get to the point wherein we can begin the conversation. Where, instead of treating the faceless like monsters or (worse) like they are invisible, either dead or (worst of all) outside the bounds of human life, we understand that they are not faceless at all. And then we can choose to not kill them (physically, symbolically, or otherwise), giving rise to the ethical relationship. And giving rise also to the self, which exists only in relation to other. But first we have to be able to see the other's face. Whatever it might look like.

And, as we've seen, the face, manipulated by transplant, by other surgery, by accident, by time, changes in multiple ways. And these manipulations may be called cosmetic or reconstructive, and these two may not be all that different. There is also another way of thinking about manipulations to the face, one that takes seriously the changes the body may effect on the mind and takes equally seriously the agency of those making those changes and choices. We have to overcome the cultural fears of evil and dominating doctors manipulating bodies for their own ends; that will only work if the patient herself is granted access, and choice, and control over her own body. We have to create an environment in which the choice to manipulate the face is not one that dramatically increases access, and choice, and advantage. We have to understand, and also ensure, that the loss of face does not mean losing face. Does not mean losing life, or entering half-life. If we take all that seriously, we won't put the faceless on trial, as we did Isabelle Dinoire, and as we do with each subsequent face-transplant recipient, testing them for their right to access this procedure. Only then, maybe, will we begin to embrace the potential of the FAT (as the press might have, and almost did, back in 2005). Face-transplant patients won't perform their trauma and normalize it in a version of makeover television in order to earn their surgeries and justify them. Unless they want to. And then, only then, may the full radical potential of the surgery, of hybridity, of new ways of conceptualizing the body, be unleashed.

CONCLUSION

I didn't solve the problem. Not even close. What I did do was point out, in this chapter and in this book, the ways that ideas about the indexicality of appearance contribute to reductive notions about identity, and the challenges that emerge when we try to overcome these forms of bodily essentialism while taking seriously lived experience and personal agency. People often work quite hard to self-fashion and project their identities in their appearance, in ways that are quite meaningful and powerful, despite the possibilities that they can be misread. And many derive a strong sense of self from the way they look and the communities with which they affiliate (voluntarily or otherwise) through their appearance and their self-fashioning, and that matters too.

It matters online, where people perform multiple versions of themselves, sometimes with greater freedom than IRL, and sometimes with a whole new set of policing constraints. It matters IRL, in terms of both the meaning we attach to our own self-presentation and (voluntary and involuntary) appearance and the meaning that others attach to it. And it matters most of all in the space between ourselves and others, and in the possibilities and limitations that emerge from the moment of encounter and interaction. That space only exists in a meaningful way when the other is seen. When the face of the other is seen, even if the face itself is absent as we traditionally define face. Or even if the face is different than it once was. The face will always be different than it once was. Difference, I suggest, is a question of degree rather than kind, no matter how drastic the change.

I can't give up on the idea that something is enjoined when we see the face of the other, even and maybe especially if the other has no face as we traditionally understand and see it. I'm willing to believe that the something (in Lévinasian terms) might be the desire to kill; surely all the face-transplant recipients that I've profiled have spoken (sometimes quite eloquently) about their experiences of social death, of people killing them in the streets. And of their responses, which include adopting the deadpan, the flat affect that Lauren Berlant describes as a strategy of personal control, an attempt to opt out of the relational system.[82] It is (always) the other who deadpans, trying to avoid being communicated with, trying to avoid being killed. That's the other side of social death: it makes the unseen (the faceless, the so-called faceless) attempt to erase their place in the system. It makes them attempt to opt out of the system of ethical relationality. Because they believe they will be killed, so they choose to remove themselves from the encounter. (And how telling the *dead* in the term, as the deadpanners make themselves be dead defensively.) Deadpan is labor, a labor that recognizes the preexisting relationality that has to be encountered or circumvented. Berlant's analysis of deadpan gives agency and control to the other who has been othered. It's a kind of a way out that depends entirely on the existence of the system itself. And I'm equally willing to believe, with Judith Butler, that as much as faces can be that which represent and even inspire humanity, they can also be that which, in the context of war propaganda, erase and deface it. But Butler's face is a concrete and nonabstract one, even as it stands in for value itself as the face of evil or the face of freedom. It can have nuance, and the nuance matters.[83] Lévinas absolutely disagrees.[84] And from them we learn: even as there is an ethical way to have face, there is no one face. There is no one way to have a face.

So let's move to a more modest proposition: Lévinas works well in the context of the (so-called) faceless as well as the faced. His idea that the other can be mastered by murder is enacted in our treatment of those whose faces we particularly can't control because they are absent. It isn't a question of not knowing, not understanding, the absent face. For Lévinas, knowledge comes second. It isn't a question of the image of the face not being visually parsed. But it is a question of response. The call to response is itself where the ethics lies, and even the faceless enjoin a response. It's just often the wrong one.

To put it another way: the index that the face provides to character has always been a construct. Except (and this is a huge exception) in so far at it links to a human being. Any human being, and one human being in particular in a particular moment. In so far as it is a manifestation of humanity, enjoining a human response and creating the space for ethical obligation and interchange.

Seriously. If we can get that far, it's very far indeed. It's not enough, but it's a start.

ACKNOWLEDGMENTS

I said: I want to put great care into the acknowledgments. It's often the only part that people read in its entirety. And a dear friend (thanked profusely below) said: sometimes you can find out exactly what kind of book it is from the acknowledgments. You can situate it in an intellectual field, in a social network, in a kind of style. Which is a generous way of saying, yes, it is the only part of the book that people read in its entirety.

It's true: the people who have helped me range widely in discipline, in method, in their relationship to me, my work, and my life. My acknowledgments underscore how interdisciplinary this work is and how deeply my ambition to think across spaces and places, to make unexpected connections and surprising alliances, permeates both my work and my life. I lie squarely in no one intellectual field or methodology; my social network is diffuse and undigital; my style is experimental and alternating. And, nevertheless, my debts are huge. My gratitude is deep. My achievements are thoroughly dependent on others. This book came out of my ongoing obsession with faces and what we imagine they can tell us. And it was forged through countless face-to-face and face-to-screen and face-to-text interactions with others, whose generosity of mind and spirit taught me the truth of the ethics of the encounter.

I am grateful for the generosity of the Dean of the Annenberg School for Communication, Michael Delli Carpini, including the teaching leave that allowed me to finish this project. Kelly Fernández and Emily Plowman are amazingly efficient and also just amazing. They make everything run more smoothly and also much more fun. The IT staff is simply extraordinary, keeping things technically smooth with not just competence but artistry.

Kyle Cassidy has been a particular visionary, making real what I have only imagined. From my first day here, Elihu Katz welcomed me with open arms, showed me the ropes, and kept me laughing the entire time. I join countless others in paying homage to him and acknowledge all that he has given to me personally and to the field as a whole. Damon Centola, Emily Falk, and Devra Moehler have been good friends across the quantitative aisle. Jessa Lingel has helped me feel at home, even though I got here first. John Jackson has been a cheerleader, an inspiration, and a role model. (If only he'd share his cloning technology with me!) And to Barbie Zelizer and Carolyn Marvin I can only say, for your mentorship, your kindness, your patience, your strength, and your wisdom—thank you. Both my experience and my work would be much impoverished without you.

The term *research assistant* does not really do justice to what these highly capable graduate students have contributed to this book. Kevin Gotkin and Aaron Shapiro helped me as I was just diving into the work; Rachel Stonecipher, Sun-Ha Hong, and Yoel Roth saw it through to its completion with creativity, diligence, and brilliance. They thought of much that I would not have, and know much that I do not, and they shared their ideas and their expertise tirelessly and generously. They, along with Piotr Szpunar, Tara Liss-Mariño, Brooke Duffy, Steven Schrag, Jonathan Pace, and Megan Genovese, have listened extensively and patiently, helping me work out and improve my ideas.

One of the tremendous advantages of the University of Pennsylvania is the excellence of the entire faculty, of which I have been a great beneficiary. Many of my colleagues here have helped me formulate this book through numerous conversations, presentations, and writing exchanges. John Tresch, Beth Linker, Ian Petric, and David Barnes have kept me anchored to my history of science and medicine roots; Peter Decherney and Tim Corrigan have helped me explore my budding interest in cinema studies with good humor and great advice; Andrea Goulet and Paul Saint-Amour have indulged and encouraged my love of literary criticism and have helped me do it better. Maria Geffen has been a steadfast friend, supporter, and source of inspiration within the university and beyond. The Center for Human Appearance here at Penn is an unusual space that has welcomed this humanist with open arms; I am particularly grateful for the support of Linton Whitaker, Sarah Kagan, Jesse Taylor, David Sarwer, and Ardis Ryder. The feminist community at Penn is strong and growing: for both personal and professional support I am deeply indebted to the wonderful women on the board of the Penn Forum for Women

Faculty, especially Susan Margulies. For helping to foster my intellectual and political commitments to gender studies I am grateful to the Gender, Sexuality, and Women's Studies program, under the extraordinary leadership of Nancy Hirschmann. Additional research funding was provided by the Trustees Council on Penn Women, which has been enormously helpful.

The final chapter of this book was written while attending the 2015 Faculty Writing Retreat; I am deeply appreciative of the organizers and the University of Pennsylvania for continuing to enable this program. Selections of this work have also been presented to audiences at the Center for Human Appearance, the History and Sociology of Science Seminar Series, the Elihu Katz colloquium at ASC, the Program on Cinema Studies, and the Faculty Conference on Disability Studies (all at the University of Pennsylvania), the colloquium series in the Communication Departments at the Hebrew University of Jerusalem and Haifa University, the Medical Humanities Program at Drexel College of Medicine, the Departments of Science Studies and of Communication at Cornell University, the annual meetings of the History of Science Society, the Society for the History of Technology, the International Communication Association, the Modern Language Association, and Poptech. I am grateful for all their feedback and suggestions.

I have been lucky in the insight, excitement, and keen critical minds of many beyond Penn. Shannon Lundeen, Adam Lowenstein, Colin Milburn, Dana Polan, Matt Stanley, Debbie Coen, Michael Gordin, Matt Rubery, Simon Goldhill, and Aviva Briefel have lent rich interdisciplinary expertise and continually remind me that thinking is and should always be fun; Katherine Sender, Zohar Kampf, Paul Frosh, David Park, Jefferson Pooley, and Fred Turner have made Communication a home for me, and it is in large part because of them that I so cherish it. John Durham Peters has been not only a role model but a mensch; Ben Peters has been a counterpart both in writing and in helping me to remember the importance of other things besides writing. Alison Winter, an extraordinary scholar, teacher, and mentor, died while I was completing this book. As with my first book, her spirit, vision, and creativity helped shape it. Molly Tolsky has given me space to again explore the joy of the written word, and John Maeda, Leetha Filderman, Kiley Lambert, and Maria Popova have reminded me of the pleasure of the spoken one. This book has flourished under the talented stewardship of Karen Merikangas Darling at the University of Chicago Press; I stand in a long line of grateful authors who pay homage to her wisdom and intellectual courage. I am also appreciative of the assistance of Evan White and the extremely helpful feedback of the two anonymous reviewers for this book. The wonderful index is courtesy of June Sawyers, who was a delight to work with.

Many (many) generous ears have heard some of these words; few (even more generous) eyes have seen them. For their enormously helpful chapter feedback and for keeping me reined in and on the right track, I am grateful to Shiamin Kwa Roses, Mary Zaborskis, Orit Halpern, and (with apologies to) Peter Lunt. Amit Pinchevski has not only heard and read but truly helped shape this project. He has continually reminded me of the larger ethics at stake in both his words and his actions, and I feel extremely lucky to have had his guidance. And Suman Seth has heroically read the entire manuscript in various pieces, applying his unparalleled critical eye, unsurpassed intellectual rigor, and unending cheerleading throughout. This is much, much better for his guidance, support, and advice.

Adina Isenberg, Claire Raab, Leora Eisenstadt, Leah Weiss, Yoella Epstein: you are all strong women whose support and love is a big part of my own strength. And to my village: Natalie and Dan, Jen and Stephan, Susan and Anthony, Puja and Chas, Todd and Stephanie, Joey and John, Tal and Yis, it definitely did take you to make it possible, and also a lot of fun (even if sleep exhaustion means I can't remember all of it). Thank you—and your children—for everything.

As this book took shape and grew, so did my own family. But it was never small: my mother Susan Pearl is a giant in deeds, in ability, in strength, and in generosity. My brother Mayeer and his wife Stacey, and their kids Reese, Nathaniel, and Harley will always be, for me, a home, no matter where life takes them (and us). My in-laws Marlene and Gerald Knepler, their daughter Miriam and her husband Avi, and Michali, Yonit, Revial, Shani, and Elinoam help make distances disappear; they keep us close even as they are far away. And to my own, most precious and immediate family, right beside me in space and closest to my heart: you trace and define the shape of my world. I am proud, I am grateful, and I am—continually and yet anew—in awe. Aria Sarelle, Melilla Eva, Yishai Eliezer—I love you. And I like you. And Ben Knepler, without whom nothing would be possible—I love you. And I like you. And I am so, so thankful for you, every day.

Our sages teach in the Ethics of Our Ancestors 4:20 that surfaces do not always tell us what lies within, and that we must be willing to look inside in order to know. But this is no simple matter, as I've outlined in this book: it takes courage, and vision, and not a little bit of love. I leave you with their words, and with a wish for all of us not only to look at what vessels contain but to understand that the contents and the vessel itself are always changing, and as we interact with them, we too change. And so do they. "Said Rabbi Meir: Look not at the vessel, but at what it contains. There are new vessels that are filled with old wine, and old vessels that do not even contain new wine."

רבי מאיר אומר, אל תסתכל בקנקן אלא במה שיש בו.
יש קנקן חדש מלא ישן, וישן שאפילו חדש אין בו.

APPENDIX: FACE TRANSPLANT
RECIPIENTS

I struggled with the question of whether to include images of the recipients alongside my text. These figures are not central to the arguments themselves and so may be construed as gratuitous. On the other hand, the visual is so fundamental to my approach that the absence of images is itself a kind of distraction, particularly given the wide availability of these representations. I decided to include these pictures as an appendix to offer the information to those who felt it necessary, while keeping open the option to consider the argument without images themselves. What follows below are pictures of the recipients prior to the damage to their faces, following their injuries, and after their transplants.

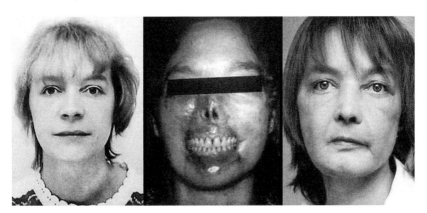

FIGURE 1. Isabelle Dinoire

188 *Appendix*

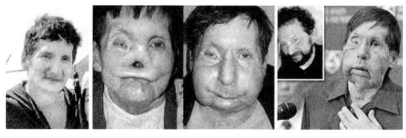

FIGURE 2. James Maki, with an insert of the face of his donor

 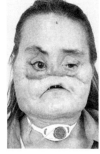  

FIGURE 3. Connie Culp

 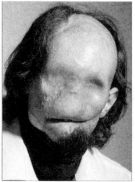 

FIGURE 4. Dallas Weins

Appendix 189

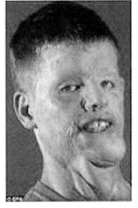
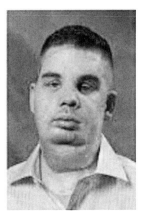

FIGURE 5. Mitch Hunter

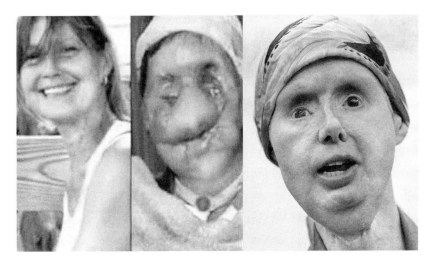

FIGURE 6. Charla Nash

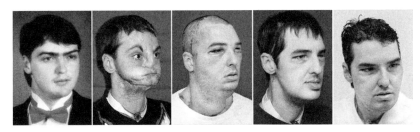

FIGURE 7. Richard Lee Norris

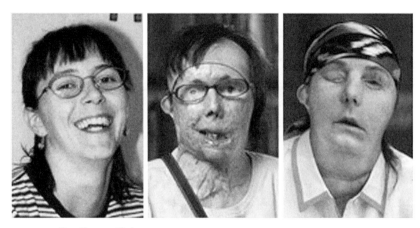

FIGURE 8. Carmen Tarleton

NOTES

CHAPTER 1

1. David Lodge, *Changing Places: A Tale of Two Campuses*, 2nd ed. (New York: Penguin Books, 1979).

2. Emmanuel Lévinas and Philippe Nemo, *Ethics and Infinity* (Duquesne University Press, 1985); Gilles Deleuze and Felix Guattari, *A Thousand Plateaus: Capitalism and Schizophrenia*, trans. Brian Massumi (Minneapolis: University of Minnesota Press, 1987); Hannah Arendt, *On Revolution* (London: Penguin Books, 1990); Erving Goffman, *The Presentation of Self in Everyday Life* (New York: Anchor, 1959); Richard Sennett, *The Fall of Public Man* (New York: WW Norton, 1992); Judith Butler, *Precarious Life: The Powers of Mourning and Violence* (London: Verso, 2004).

3. Sharrona Pearl, *About Faces: Physiognomy in Nineteenth-Century Britain* (Cambridge, MA: Harvard University Press, 2010).

4. I borrow here from Donna Haraway's famous formulation in Donna Jeanne Haraway, "A Cyborg Manifesto: Science, Technology, and Socialist-Feminism in the Late Twentieth Century," in *Simians, Cyborgs, and Women: The Reinvention of Nature* (New York: Routledge, 1991), 149–81.

5. Sharrona Pearl, "Believing in Not Seeing: Teaching Atrocity without Images," *Afterimage* 40, no. 6 (2013): 16–20; Sharrona Pearl and Alexandra Sastre, "The Image Is (Not) the Event: Negotiating the Pedagogy of Controversial Images," *Visual Communication Quarterly* 21, no. 4 (2014): 198–209, doi:10.1080/15551393.2014.987283; Sharrona Pearl, "Victorian Blockbuster Bodies and the Freakish Pleasure of Looking," *Nineteenth-Century Contexts* 38 (forthcoming).

CHAPTER 2

1. There are surprisingly few reports on the Sandeep Kaur case, though her young age at the time of the surgery may be a factor in the lack of coverage. There was some renewed interest in Kaur following the first facial allograft in 2005, but coverage referenced her case as

a historical precedent rather than an interesting case in its own right. Early reports of the Kaur story are framed as "gruesome" and "horrific," and little attention is paid to the medical details. The most comprehensive coverage of Kaur can be found in a 2003 documentary *Human Face Transplant* directed by Eains Colley. The film aired on the British station Channel 4 in 2004 as part of the media attention around the possibility of face transplants becoming viable in the near future.

2. A Pubmed search found no articles dealing with the technicalities of the Sandeep Kaur case in depth, though it was mentioned in passing in subsequent articles dealing with facial allografts.

3. The technicalities of the Kaur face transplant were similar to those of a hand transplant, which, at the time, had not yet been successfully performed.

4. While this is still an alignment narrative akin to those deployed in makeover contexts, the fact that Kaur attempted to regain her former self rather than create a new exterior appearance is a significant difference. Her young age was also a factor in protecting her from criticism for being subject to vanity in undertaking a possibly unnecessary medical risk.

5. I explore this trope in depth later in the book.

6. As a minor and an unconscious patient, Kaur's informed consent was provided by her mother. This kind of emergent case requiring innovative and untested approaches has received a great deal of treatment in the bioethical literature. There is broad consensus that emergency situations are governed by somewhat different sets of considerations given the time-sensitive nature of the decisions that are made. While Kaur's doctor did not have ethical approval for this procedure *as such*, he was empowered to make the decision to proceed in the best interests of the patient given all factors in the situation. For more on emergency-physician ethics, see "Code of Ethics for Emergency Physicians," American College of Emergency Physicians, accessed July 13, 2015, http://www.acep.org/Clinical---Practice-Management/Code-of-Ethics-for-Emergency-Physicians/.

7. Cleft palate surgery is routine when resources exist to cover the cost. Untreated conditions can lead to feeding and speaking difficulties, repeated ear infections, and social stigma that, in certain cases and places, can lead to severe and even life-threatening mistreatment. For more on the medical approach to cleft palate, see "Facts about Cleft Lip and Cleft Palate," *Centers for Disease Control and Prevention*, April 12, 2015, http://www.cdc.gov/ncbddd/birthdefects/CleftLip.html.

8. I refer in particular to parents of children with Down's Syndrome who opt for cosmetic surgery to minimize its visible markers. The case of young Georgia Bussey sparked a national conversation on the topic in England in 1998 following the ITV documentary *Changing Faces* that aired their story. See Nick Finnis, *Changing Faces* (United Kingdom: ITV Studios Global Entertainment, 1998), https://itvstudios.com/programmes/changing-faces. For more on this case, see "Down's Syndrome Mother Denies Vanity," *BBC News*, November 18, 1998, http://news.bbc.co.uk/2/hi/health/216479.stm. Numerous bioethicists have discussed this issue over time; there is broad consensus that adults with Down's have the right to choose surgery, but the question is far more fraught when it comes to children.

9. Abraham Thomas et al., "Total Face and Scalp Replantation," *Plastic and Reconstructive Surgery* 102, no. 6 (1998): 2085–87.

10. "Woman's Face Reattached in Rare Surgery," *Seattle Times*, September 22, 1997, http://community.seattletimes.nwsource.com/archive/?date=19970922&slug=2561892.

11. The entire history of medicine and surgery is, in its way, the history of experimentation; as I discuss later in this chapter, that is certainly true in the history of organ transplantation.

12. Margrit Shildrick has discussed the Cartesian overtones to cosmetic surgery and the ways that the psychic consequences of physical manipulations are elided. Margrit Shildrick, "Corporeal Cuts: Surgery and the Psycho-Social," *Body & Society* 14, no. 1 (2008): 31–46, doi:10.1177/1357034X07087529. Debra Gimlin has argued that plastic surgery can be a way to limit the incursion of an (undesirable) body into one's consciousness. Surgical manipulation then becomes a coping strategy that minimizes concern with, and thoughts about, the body. Debra Gimlin, "Imagining the Other in Cosmetic Surgery," *Body & Society* 16, no. 4 (2010): 57–76, doi:10.1177/1357034X10383881.

13. Carla Bluhm and Nathan Clendenin, *Someone Else's Face in the Mirror: Identity and the New Science of Face Transplants* (Westport, CT: Praeger, 2009).

14. In this ambition, I stand on the shoulders of other scholars. For a recent take on this vision, see posthuman disability-studies scholar Dan Goodley, *Disability Studies: An Interdisciplinary Introduction* (London: Sage, 2011). While Goodley's analysis tends toward a utopian technological futurism that does not provide enough space for the individual subject, he provides a space of recuperation about human difference and the ways it can be reimagined. See also Donna Reeve, "Cyborgs, Cripples and iCrip: Reflections on the Contribution of Haraway to Disability Studies," in *Disability and Social Theory*, ed. Dan Goodley, Bill Hughes, and Lennard Davis (New York: Palgrave Macmillan, 2012), 91–111.

15. Carolyn Marvin, "Communication as Embodiment," in *Communication As . . . : Perspectives on Theory*, ed. Gregory J. Shepherd, Jeffrey St. John, and Theodore G. Striphas (Thousand Oaks, CA: Sage, 2006), 67–74. For the storytelling potentiality of the body, see Edmund Coleman-Fountain and Janice McLaughlin, "The Interactions of Disability and Impairment," *Social Theory & Health* 11, no. 2 (2013): 133–50, doi:10.1057/sth.2012.21. For the communicative aspects of physical pain in particular, see Tobin Siebers, *Disability Aesthetics*, Corporealities: Discourses of Disability (Ann Arbor: University of Michigan Press, 2010).

16. Allan Sekula, "The Body and the Archive," *October* 39 (1986): 3–64. As many scholars have pointed out, the archive is hardly a neutral or objective space but, rather, is the result of multiple sorting and curating mechanisms that are themselves reflective of social and institutional power hierarchies. For the classic text on this, see Jacques Derrida, *Archive Fever: A Freudian Impression* (Chicago: University of Chicago Press, 1996).

17. Colleen Farrell, "Telltale Purple Lesions," *Atrium: The Report of the Northwestern Medical Humanities and Bioethics Program* 9 (Spring 2011): 10–11.

18. Sander L. Gilman, *Making the Body Beautiful: A Cultural History of Aesthetic Surgery* (Princeton, NJ: Princeton University Press, 1999).

19. Susan M. Schweik, *The Ugly Laws: Disability in Public* (New York: New York University Press, 2010).

20. Meghan Daum, "In-Your-Face Journalism," *Los Angeles Times*, December 3, 2005.

21. Haraway, "A Cyborg Manifesto."

22. Ibid.; Susan Bordo, *Unbearable Weight: Feminism, Western Culture, and the Body*

(Berkeley: University of California Press, 2004); Rosemarie Garland-Thomson, "Integrating Disability, Transforming Feminist Theory," *NWSA Journal* 14, no. 3 (2002): 1–32.

23. Gilman, *Making the Body Beautiful*.

24. Annette Drew-Bear, *Painted Faces on the Renaissance Stage: The Moral Significance of Face-Painting Conventions* (Lewisburg, PA: Bucknell University Press, 1994).

25. Gilman, *Making the Body Beautiful*; Kathy Lee Peiss, *Hope in a Jar: The Making of America's Beauty Culture* (Philadelphia: University of Pennsylvania Press, 2011).

26. Bernadette Wegenstein and Nora Ruck, "Physiognomy, Reality Television and the Cosmetic Gaze," *Body & Society* 17, no. 4 (2011): 46, doi:10.1177/1357034X11410455.

27. Sander L. Gilman, *Creating Beauty to Cure the Soul: Race and Psychology in the Shaping of Aesthetic Surgery* (Durham, NC: Duke University Press, 1998).

28. Diane Naugler, "Crossing the Cosmetic/Reconstructive Divide: The Instructive Situation of Breast Reduction Surgery," in *Cosmetic Surgery: A Feminist Primer*, ed. Cressida J. Heyes and Meredith Jones (Farnham: Ashgate, 2009), 224–37. See also Meredith Jones, *Skintight: An Anatomy of Cosmetic Surgery* (New York: Berg, 2008), 3. Jones points out that plastic and cosmetic surgery procedures are often identical save the values attached to each. Anne Balsamo lays out these distinctions in Anne Balsamo, "On the Cutting Edge: Cosmetic Surgery and the Technological Production of the Gendered Body," *Camera Obscura* 10, no. 128 (1992): 206–37. The television show *Nip/Tuck* explores this division in depth from its first episode through the entire series. John Liesch and Cindy Patton offer the example of gay men using facial fillers as a treatment for the HIV/AIDS symptom of facial wasting, again blurring the distinction between cosmetic and reconstructive. John Liesch and Cindy Patton, "Clinic or Spa? Facial Surgery in the Context of AIDS-Related Facial Wasting," in *The Rebirth of the Clinic*, ed. Cindy Patton (Minneapolis: University of Minnesota Press, 2010), 1–16.

29. Kathy Davis, *Reshaping the Female Body: The Dilemma of Cosmetic Surgery* (New York: Routledge, 1995).

30. For more on these debates, see Peter D. Kramer, *Listening to Prozac: A Psychiatrist Explores Antidepressant Drugs and the Remaking of the Self* (New York: Viking Press, 1993).

31. For more on the effects of antirejection medication in face transplants, see Shehan Hettiaratchy et al., "Composite Tissue Allotransplantation—A New Era in Plastic Surgery?" *British Journal of Plastic Surgery* 57, no. 5 (2004): 381–91, doi:10.1016/j.bjps.2004.02.012.

32. Sarah Kember and Joanna Zylinska, *Life After New Media: Mediation as a Vital Process* (Cambridge, MA: MIT Press, 2012), 151.

33. Wegenstein and Ruck, "Physiognomy, Reality Television and the Cosmetic Gaze."

34. Katherine Sender, *The Makeover: Reality Television and Reflexive Audiences*, Critical Cultural Communication (New York: New York University Press, 2012), 7.

35. Heather Laine Talley, *Saving Face: Disfigurement and the Politics of Appearance* (New York: New York University Press, 2014), 150.

36. For more on neoliberalism, see Philip Mirowski and Dieter Plehwe, eds., *The Road from Mont Pèlerin: The Making of the Neoliberal Thought Collective* (Cambridge, MA: Harvard University Press, 2009).

37. See for example Coleman-Fountain and McLaughlin, "The Interactions of Disability and Impairment."

38. Thomas Abrams, "Being-towards-Death and Taxes: Heidegger, Disability and the Ontological Difference," *Canadian Journal of Disability Studies* 2, no. 1 (2013): 28–50.

39. "Special Issue on Face Transplantation," *American Journal of Bioethics* 4, no. 3 (2004).

40. These discussions dominate the *AJOB* special issue; the question of accommodation borrows from disability-studies literature and is explored in this chapter.

41. Talley, *Saving Face*, 5.

42. This definition of disability—itself a fraught and highly debated term—is taken from the United Nations definition: "Any restriction or lack (resulting from an impairment) of ability to perform an activity in the manner or within the range considered normal for a human being." For more on the various definitions of the term and the politics behind them, see Deborah Kaplan, "The Definition of Disability: Perspective of the Disability Community," *Journal of Health Care Law and Policy* 3, no. 2 (2000): 352–64.

43. Goffman, *The Presentation of Self*.

44. The critique of narratives of overcoming was most influentially outlined in Simi Linton, *Claiming Disability: Knowledge and Identity* (New York: New York University Press, 1998).

45. Garland-Thomson, "Integrating Disability, Transforming Feminist Theory"; Garland-Thomson, "Feminist Disability Studies," *Signs* 30, no. 2 (2005): 1557–87.

46. Garland-Thomson, "Integrating Disability, Transforming Feminist Theory."

47. By placing Garland-Thomson and Gilman in conversation, I seek to heed Beth Linker's analytically powerful call to more effectively integrate medical history and disability studies. Beth Linker, "On the Borderland of Medical and Disability History: A Survey of the Fields," *Bulletin of the History of Medicine* 87, no. 4 (2013): 499–535, doi:10.1353/bhm.2013.0074.

48. Disability studies deals with facial disfigurement in a limited way, and with facelessness in an even more peripheral fashion, so much of so much of my positioning of facial-allograft procedures in the discourse of disability is extrapolation from general arguments.

49. Marjorie Kruvand explores the power of medical rhetoric in dictating the conversation around face transplants in particular. See Marjorie Kruvand, "Face to Face: How the Cleveland Clinic Managed Media Relations for the First U.S. Face Transplant," *Public Relations Review* 36, no. 4 (2010): 367–75, doi:10.1016/j.pubrev.2010.09.001.

50. April Herndon, "Disparate but Disabled: Fat Embodiment and Disability Studies," *NWSA Journal* 14, no. 3 (2002): 120–37. Herndon's formulation insists on fatness as a communal categorization; I resist this approach in favor of a more subjective narrative but still usefully draw on her insights around medicalization.

51. For a discussion of the neoliberal and moralizing overtone of fatness in *The Biggest Loser*, see Michael L. Silk, Jessica Fancombe, and Faye Bachelor, "The Biggest Loser: The Discursive Constitution of Fatness," *Interactions: Studies in Communication & Culture* 1, no. 3 (2011): 369–89, doi:10.1386/iscc.1.3.369_1.

52. Radhika Parameswaran, "Global Queens, National Celebrities: Tales of Feminine Triumph in Post-liberalization India," *Critical Studies in Media Communication* 21, no. 4 (2004): 346–70, doi:10.1080/0739318042000245363.

53. Imani Perry, "Buying White Beauty," *Cardozo Journal of Law & Gender* 12 (Spring 2006): 579–607.

54. For a postfeminist critique of this approach, as well as a comprehensive survey on

the literature, see Ruth Holliday and Jacqueline Sanchez Taylor, "Aesthetic Surgery as False Beauty," *Feminist Theory* 7, no. 2 (2006): 179–95, doi:10.1177/1464700106064418. For a more ethnographic approach, see Gimlin, "Imagining the Other in Cosmetic Surgery."

55. For a quantitative approach to the value of attractiveness, see Daniel S. Hamermesh, *Beauty Pays: Why Attractive People Are More Successful* (Princeton, NJ: Princeton University Press, 2011).

56. Linda Colley, *Britons: Forging the Nation, 1707–1837*, rev. ed. (New Haven, CT: Yale University Press, 2009).

57. Numerous articles and editorials about Zellweger appeared both in the tabloids and in the broadsheets. See for example Alex Kuczynski, "Why the Strong Reaction to Renée Zellweger's Face?" *New York Times*, October 24, 2014, http://www.nytimes.com/2014/10/26/fashion/why-the-strong-reaction-to-renee-zellweger-face.html. Uma Thurman claimed that her altered appearance was due to an experimental makeup technique, which she discussed during an appearance on *Today* on February 12, 2015. The interview can be viewed at "Uma Thurman: Physical Discipline of Kids 'Inhumane,'" *Today*, NBC, February 12, 2015, http://www.today.com/video/today/56965916#56965916.

58. Sennett, *The Fall of Public Man*.

59. For the rise of the heroic medical narratives, see José van Dijck, *The Transparent Body: A Cultural Analysis of Medical Imaging* (Seattle: University of Washington Press, 2005). I take up the theme of transparency of plastic-surgery interventions later in this book.

60. David B. Sarwer and Canice E. Crerand, "Body Dysmorphic Disorder and Appearance Enhancing Medical Treatments," *Body Image* 5, no. 1 (2008): 50–58, doi:10.1016/j.bodyim.2007.08.003.

61. Gimlin, "Imagining the Other in Cosmetic Surgery." For an overview of the psychological risks of plastic surgery, see Melissa Dittmann, "Plastic Surgery: Beauty or Beast?" *Monitor on Psychology* 36, no. 8 (2005): 30. Former *Swan* contestant Lorrie Arias has spoken out volubly about the damage the experience of being on the show, and the surgeries she received, did to her psychologically and physically. See "Former 'Swan' Contestant Says Erin Moran Will Need Major Therapy before Appearing on 'Celebrity Swan,'" *Huffington Post*, February 22, 2013, http://www.huffingtonpost.com/2013/02/22/celebrity-swan-erin-moran_n_2741082.html

62. Dittmann offers examples of both kinds of outcomes in her overview of the psychological impact of extreme plastic surgery.

63. Kevin O'Leary, "Inside Renée Zellweger's Shocking Decision to Have Plastic Surgery," *Boston Herald*, November 3, 2014.

64. For the role insurance plays in disease categories, see Allan Young, *The Harmony of Illusions: Inventing Post-Traumatic Stress Disorder* (Princeton, NJ: Princeton University Press, 1997).

65. I discuss the debates around the medical necessity of the facial allograft in great depth in subsequent chapters of this book.

66. For a detailed history of plastic surgery, see Elizabeth Haiken, *Venus Envy: A History of Cosmetic Surgery* (Baltimore: Johns Hopkins University Press, 1997).

67. Victoria Pitts-Taylor, *Surgery Junkies: Wellness and Pathology in Cosmetic Culture* (New Brunswick, NJ: Rutgers University Press, 2007).

68. Diane Von Furstenberg, "Diane Von Furstenberg Shares Her Philosophy on Aging in an Excerpt from Her New Memoir," *Vogue*, September 7, 2014, http://www.vogue.com/1156531/diane-von-furstenberg-book-the-woman-i-wanted-to-be/.

69. There is a rich literature on prosthetics, militarization, and the rhetoric of injured masculinity that explores the relationship between war and medical developments, including Beth Linker, *War's Waste: Rehabilitation in World War I America* (Chicago: University of Chicago Press, 2011); Joanna Bourke, "The Battle of the Limbs: Amputation, Artificial Limbs and the Great War in Australia," *Australian Historical Studies* 29, no. 110 (1998): 49–67, doi:10.1080/10314619808596060; Lisa Herschbach, "Prosthetic Reconstructions: Making the Industry, Re-Making the Body, Modeling the Nation," *History Workshop Journal* Autumn, no. 44 (1997): 22–57, accessed July 13, 2016, http://hwj.oxfordjournals.org/content/1997/44/22.full.pdf; David Serlin, *Replaceable You: Engineering the Body in Postwar America* (Chicago: University Of Chicago Press, 2004); Vanessa Warne, "'To Invest a Cripple with Peculiar Interest': Artificial Legs and Upper-Class Amputees at Mid-Century," *Victorian Review* 35, no. 2 (2009): 83–100; Stephen Mihm, "'A Limb Which Shall Be Presentable in Polite Society': Prosthetic Technologies in the Nineteenth Century," in *Artificial Parts, Practical Lives: Modern Histories of Prosthetics* (New York: New York University Press, 2002), 282–99.

70. Carolyn Marvin, *Blood Sacrifice and the Nation: Totem Rituals and the American Flag*, Cambridge Cultural Social Studies (Cambridge: Cambridge University Press, 1999).

71. Giorgio Sperati, "Amputation of the Nose throughout History," *Acta Otorhinolaryngologica Italica* 29, no. 1 (2009): 44–50.

72. John Hall, "'Never Give Up': Afghan Teenager Who Had Nose and Ears Cut off by Abusive Husband Reveals New Life and New Face," *The Independent*, February 26, 2013, http://www.independent.co.uk/news/world/americas/never-give-up-afghan-teenager-who-had-nose-and-ears-cut-off-by-abusive-husband-reveals-new-life-and-new-face-8511958.html.

73. Sperati, "Amputation of the Nose throughout History," 44.

74. Pearl, *About Faces*.

75. Exod. 21:6, Hebrew Bible.

76. Pearl, *About Faces*. See in particular the discussion of Mary Ann Bell on p. 53.

77. Wegenstein and Ruck, "Physiognomy, Reality Television and the Cosmetic Gaze."

78. Michel Foucault draws on his notion of biopower in numerous writings; for a discussion of more contemporary applications, see Paul Rabinow and Nikolas Rose, "Biopower Today," *BioSocieties: An Interdisciplinary Journal for the Social Study of the Life Sciences* 1, no. 2 (2006): 195–217, doi:10.1017/S1745855206040014.

79. Haraway, "A Cyborg Manifesto"; Goodley, *Disability Studies*.

80. van Dijck, *The Transparent Body*.

81. The historical performers Cheng and Eng provide strong support for the possibility of living fulfilled conjoined lives, though the historical record has been highly embellished in their case. There are a number of conjoined twins who could not be separated safely; some have been quite public about the challenges they face and their desire for independence. In other cases, the separation operation itself created numerous debilitating medical conditions. For some examples of conjoined twins, see "10 Stories of Conjoined Twins," *Mental Floss*, February 27, 2014, http://mentalfloss.com/article/55329/10-stories-conjoined-twins.

82. We will explore individual examples and narratives around facelessness later in this book.

83. For trends in plastic surgery, see "Infographic: 2013 Plastic Surgery Trends," American Society of Plastic Surgeons, accessed July 13, 2015, http://www.plasticsurgery.org/news/plastic-surgery-statistics/2013/infographic-trends.html.

84. While the gap is a more recent trend, older model Lauren Hutton sported the quirk far earlier. Other notable quirks include supermodel Cindy Crawford's upper lip mole, which many agencies told her to have removed in order for her to be successful. The mole was removed from her early print work, including her first cover of *Vogue* in 1987.

85. For more on trends, see Duncan J. Watts, *Six Degrees: The Science of a Connected Age* (New York: Norton, 2003).

86. For more on digital celebrity, see Sharon Marcus, ed., "Special Issue on Celebrities and Publics in the Internet Era," *Public Culture* 27, no. 1 (2015).

87. I explore the face as mask in chapter 3.

88. Nadja Durbach, *Spectacle of Deformity: Freak Shows and Modern British Culture* (Berkeley: University of California Press, 2010).

89. Lorraine Daston and Katharine Park, *Wonders and the Order of Nature, 1150–1750* (New York: Zone Books, 1998).

90. Authenticity is fraught and highly complicated ground. See John L. Jackson, *Real Black: Adventures in Racial Sincerity* (Chicago: University of Chicago Press, 2005).

91. Margaret M. Lock, *Twice Dead: Organ Transplants and the Reinvention of Death*, California Series in Public Anthropology 1 (Berkeley: University of California Press, 2002).

92. Ibid.; Susan E. Lederer, *Flesh and Blood: Organ Transplantation and Blood Transfusion in Twentieth-Century America* (Oxford: Oxford University Press, 2008).

93. For a more detailed history, see Lock, *Twice Dead*; Lederer, *Flesh and Blood*; Ayesha Nathoo, *Hearts Exposed: Transplants and the Media in 1960s Britain*, Science, Technology, and Medicine in Modern History (Basingstoke: Palgrave Macmillan, 2009).

94. Albert R. Jonsen, *The Birth of Bioethics* (New York: Oxford University Press, 1998), 197, http://site.ebrary.com/id/10375228.

95. Ibid.

96. Firdos Alam Khan, *Biotechnology Fundamentals* (Boca Raton, FL: CRC Press, 2012), 305.

97. Jonsen, *The Birth of Bioethics*, 197.

98. Gerald Leach, *The Biocrats: Implications of Medical Progress* (London: Cape, 1970), 286.

99. The Latin version of this phrase, *primum non nocere*, is widely taught to health and medical professionals. Its origin is disputed, but the Hippocratic Oath does require a promise "to abstain from doing harm."

100. Lederer explores the question of blood-transfusion refusal in depth. See especially Lederer, *Flesh and Blood*, 192–97. The case of K'aila Paulette highlights the question of parental rights. Paulette's parents made the decision not to place their infant son on a list for a liver transplant for his end-stage renal disease. Despite the risky prognosis for transplants in infants, Paulette's doctors contacted child services and attempted to remove the child from his parents. For details of this case, see Lock, *Twice Dead*, 310–15.

101. Jonsen, *The Birth of Bioethics*, 198.

102. Marcel Mauss, *The Gift: The Form and Reason for Exchange in Archaic Societies*, trans. W. D. Halls (New York: WW Norton, 1990).

103. Arjun Appadurai, "Introduction: Commodities and the Politics of Value," in *The Social Life of Things: Commodities in Cultural Perspective*, ed. Arjun Appadurai (Cambridge: Cambridge University Press, 1986), 3–63; Mary Douglas, "No Free Gifts," foreword to *The Gift: The Form and Reason for Exchange in Archaic Societies*, by Marcel Mauss, trans. W. D. Halls (New York: WW Norton, 1990); Annette B. Weiner, *Inalienable Possessions: The Paradox of Keeping While Giving* (Berkeley: University of California Press, 1990).

104. For more on legal developments, see Jonsen, *The Birth of Bioethics*, 204.

105. Christiaan Barnard was on the cover of *Time* on December 15, 1967.

106. And many did; there was no shortage of patients willing to try the transplant procedure in its early days.

107. In bioethical terms, willingness to engage in experimental procedures for research purposes rather than as a potential treatment is one of the ways to deal with the problem of the impossibility of informed consent in desperate cases. I discuss this approach later in the book.

108. The principle of beneficence was applied to bioethics from the late 1970s. Tom Beauchamp, "The Principle of Beneficence in Applied Ethics," in *Stanford Encyclopedia of Philosophy*, ed. Edward N. Zalta, October 3, 2013, http://plato.stanford.edu/archives/win2013/entries/principle-beneficence/.

109. Shana Alexander, "They Decide Who Lives, Who Dies: Medical Miracle Puts Moral Burden on Small Committee," *Life*, November 9, 1962.

110. For an overview of these suggestions, see Jonsen, *The Birth of Bioethics*, 217.

111. Ibid., 218.

112. Ad Hoc Committee of the Harvard Medical School to Examine the Definition of Death. "A Definition of Irreversible Coma," *Journal of the American Medical Association* 205, no. 6 (1968): 85–88.

113. For details of the various forms this committee has taken, see "History of Bioethics Committees," Presidential Commission for the Study of Bioethical Issues, accessed July 13, 2015, http://bioethics.gov/history.

114. Daniel Callahan, "Bioethics as a Discipline," *Hastings Center Studies* 1 (1973): 66–73.

115. The strongest advocate of cellular memory is Dr. Paul Pearsell. A collection of his research can be found on his website. Paul Pearsell, Gary E. Schwartz, and Linda G. Russek, "Organ Transplants and Cellular Memories," April 12, 2015, http://www.paulpearsall.com/info/press/3.html.

116. See Lock, *Twice Dead*, 320.

117. Claire Sylvia, "'I Was Given a Young Man's Heart—and Started Craving Beer and Kentucky Fried Chicken,' Says 47-Year-Old Mum," *Daily Mail*, April 9, 2008. See also the story of heart-transplant recipient David Waters, recounted by Richard Shears. Richard Shears, "Do Hearts Have Memories? Transplant Patient Gets Craving for Food Eaten by Organ Donor," *Daily Mail*, December 23, 2009.

118. Lesley A. Sharp, "Organ Transplantation as a Transformative Experience: Anthropological Insights into the Restructuring of the Self," *Medical Anthropology Quarterly* 9, no. 3 (1995): 357–89, doi:10.1525/maq.1995.9.3.02a00050.

119. Lock, *Twice Dead*, 323.

120. Ibid.

121. Isabelle Dinoire continued to think of her new face as belonging to someone else; Marc LaFrance has argued that her resistance was part of a larger set of psychological issues and challenges that she faced. Marc LaFrance, "'She Exists within Me': Subjectivity, Embodiment, and the World's First Face Transplant," in *Abjectly Boundless: Boundaries, Bodies and Health Work*, ed. Trudy Rudge and Dave Holmes (Farnham: Ashgate, 2010), 147–59.

122. Sharp, "Organ Transplantation as a Transformative Experience."

123. Lederer, *Flesh and Blood*, x.

124. Edgardo D. Carosella and Thomas Pradeu, "Transplantation and Identity: A Dangerous Split?" *Lancet* 368, no. 9531 (2006): 183–84.

125. Ibid., 183.

126. "Cape Coloured" was a specific racial designation in apartheid South Africa for those of mixed race.

127. "The Telltale Heart," *Ebony*, March 1968.

128. Ibid.

129. Gurvinder G. Kalra and Dinesh Bhugra, "Representation of Organ Transplantation in Cinema and Television," *International Journal of Organ Transplantation Medicine* 2, no. 2 (2011): 93–100; Robert D. O'Neill, "'Frankenstein to Futurism': Representations of Organ Donation and Transplantation in Popular Culture," *Transplantation Reviews* 20, no. 4 (2006): 222–30, doi:10.1016/j.trre.2006.09.002.

130. Mark Dorrian, "On the Monstrous and the Grotesque," *Word & Image* 16, no. 3 (2000): 310–17, doi:10.1080/02666286.2000.10435688.

131. This phenomenon should not be confused with "alien hand" or "anarchic hand," a neurobiological condition describing hand movements performed against the patient's will, like the self-strangulation scene of the titular character in *Dr. Strangelove*. The alien-hand symptom is due to brain lesions and not transplantation. For more on the phenomenon, see Clelia Marchetti and Sergio Della Sala, "Disentangling the Alien and Anarchic Hand," *Cognitive Neuropsychiatry* 3, no. 3 (1998): 191–207, doi:10.1080/135468098396143.

132. Patrick Gonder, "Like a Monstrous Jigsaw Puzzle: Genetics and Race in Horror Films of the 1950s," *The Velvet Light Trap* 52, no. 1 (2003): 35–36, doi:10.1353/vlt.2003.0022.

133. Kelly Hurley, "Reading Like an Alien: Posthuman Identity in Ridley Scott's *Alien* and David Cronenberg's *Rabid*," in *Posthuman Bodies*, ed. Judith Halberstam and Ira Livingston (Bloomington: Indiana University Press, 1995), 203.

134. Gonder, "Like a Monstrous Jigsaw Puzzle," 33.

135. Ian Olney, "The Problem Body Politic, or 'These Hands Have a Mind All Their Own!': Figuring Disability in the Horror Film Adaptations of Renard's *Les Mains d'Orlac*," *Literature/Film Quarterly* 34, no. 4 (2006): 294–302.

136. Lennard J. Davis, "Constructing Normalcy," in *The Disability Studies Reader*, ed. Lennard J. Davis, 3rd ed. (New York: Routledge, 2010), 9.

137. Olney, "The Problem Body Politic," 301.

138. Stephanie Brown Clark, "Frankenflicks: Medicine and Monstrosity in Fiction Film," in *Cultural Sutures: Medicine and Media*, ed. Lester D. Friedman (Durham, NC: Duke University Press, 2004), 132.

139. Ibid., 147.
140. Ibid., 148.
141. I have not addressed the central question of the donor in this chapter; that comes later.
142. And it does go wrong sometimes. There is, for example, an entire Pinterest board devoted to "plastic surgery gone wrong": Dalia Morgan, "Plastic Surgery Gone Wrong," April 12, 2015, https://www.pinterest.com/morgandalia/plastic-surgery-gone-wrong/.
143. Mark A. Rembis, "Disability Studies," in *International Encyclopedia of Rehabilitation*, ed. J. H. Stone and Maurice Blouin, 2010, http://cirrie.buffalo.edu/encyclopedia/en/article/281/.
144. Oscar Wilde, *The Picture of Dorian Gray*, Dover Thrift Editions (New York: Dover Publications, 1993).

CHAPTER 3

1. For more on the reorganization of media and sense, see Orit Halpern, *Beautiful Data: A History of Vision and Reason Since 1945*, Experimental Futures (Durham, NC: Duke University Press, 2014); Fred Turner, *The Democratic Surround: Multimedia & American Liberalism from World War II to the Psychedelic Sixties* (Chicago: University of Chicago Press, 2013).

2. The scholarship on transplant films has focused largely on the effect these films have on people's attitudes toward organ donation. For a listing of these films and a review of their effects, see Kalra and Bhugra, "Representation of Organ Transplantation in Cinema and Television." For a more taxonomic approach that explores changes in the tropes in these films over time, see O'Neill, "'Frankenstein to Futurism.'" O'Neill argues for a difference between the representation of bodily mutilation in the classic horror genre and the contemporary postmodern approach.

3. *Eyes without a Face* deals most directly with bodily mutilation as such, drawing on the gothic legacy of *Frankenstein* with its hidden laboratory of horror set in a gloomy, isolated mansion. The other two films are more indirectly horrible, playing with the dark capabilities and emotional desperation of humanity as its form of mutilation. While bodies are manipulated in *Seconds* and *The Face of Another*, it is souls that are more clearly mutilated.

4. Vivian Carol Sobchack, *Screening Space: The American Science Fiction Film*, 2nd ed. (New Brunswick, NJ: Rutgers University Press, 1997).

5. Christine Cornea, "Frankenheimer and the Science Fiction/Horror Film," in *A Little Solitaire: John Frankenheimer and American Film*, ed. Murray Pomerance and R. Barton Palmer (New Brunswick, NJ: Rutgers University Press, 2011), 230.

6. Hurley, "Reading Like an Alien," 203.

7. Paul K. Longmore, "Screening Stereotypes: Images of Disabled People," in *Screening Disability: Essays on Cinema and Disability*, ed. Christopher E. Smit and Anthony Enns (Lanham, MD: University Press of America, 2001), 1–17.

8. There are some blurbs for *The Face of Another* that mention the other two films, including TCM's brief overview of the title. "The Face of Another," Turner Classic Movies, July 24, 2014, http://www.tcm.com/this-month/article/150490|0|The-Face-of-Another.html. The Offscreen Film Festival in Belgium screened the three films together in February 2011.

9. There are of course scores of films dealing with changing how one looks, both in the sci/

fi and horror genres as well as in drama, romantic comedy, and numerous other forms. This broad categorization would include films that deal with plastic surgery, makeup, scarification, tattooing, and numerous other ways of changing one's appearance. Even within the narrower category of bodily mutilation and horror, there are too many films to list. Later in this chapter, I briefly discuss an even more specific vector, films about facial manipulation, in order to show why even that characterization is too broad for my analytic purposes.

10. George Franju was an important predecessor to the New Wave movement.

11. Adam Lowenstein, *Shocking Representation: Historical Trauma, National Cinema, and the Modern Horror Film*, Film and Culture (New York: Columbia University Press, 2005).

12. Ibid., 40.

13. Stefanos Geroulanos, "Postwar Facial Reconstruction: Georges Franju's *Eyes without a Face*," *French Politics, Culture & Society* 31, no. 2 (2013): 25.

14. There is a great deal of secondary literature on both *Seconds* and *Eyes without a Face*. *The Face of Another*, widely regarded as the least successful of Teshigahara's trilogy, is less well examined, though it does command some attention. *Eyes without a Face* is, according to the Criterion blurb, "a major influence on the genre in the decades since its release." See details online at *"Eyes without a Face*," Criterion Collection, July 24, 2014, http://www.criterion.com/films/950-eyes-without-a-face.

15. Isabel Quigley, "The Small Savages," *Spectator*, February 5, 1960, 182.

16. Dilys Powell, "A Fit of the Horrors," *Sunday Times* (London), June 16, 1963.

17. David Kalat, *"Eyes without a Face:* The Unreal Reality," Criterion Collection, October 16, 2013, https://www.criterion.com/current/posts/343-eyes-without-a-face-the-unreal-reality.

18. Pauline Kael, *5001 Nights at the Movies: A Guide from A to Z* (New York: Holt, Rinehart and Winston, 1982), 230.

19. Kalat, *"Eyes without a Face."*

20. The more contemporary positive reviews are too numerous to list, but there is a good compilation at: *"Eyes without a Face*: External Reviews," *IMDb*, April 12, 2015, http://www.imdb.com/title/tt0053459/externalreviews?ref_=tt_ql_op_5. For a discussion of the cinematic influence of the film on both horror and art-house genres, see Joan Hawkins, *Cutting Edge: Art-Horror and the Horrific Avant-Garde* (Minneapolis: University of Minnesota Press, 2000), 65–66.

21. Gerald Pratley, *The Films of Frankenheimer: Forty Years in Film* (Bethlehem, PA: Lehigh University Press, 1998), 56; James D. Powers, "Review of *Seconds*," *The Hollywood Reporter*, March 26, 1962, 56; David Wilson, "Review of *Seconds*," *Sight and Sound* 36, no. 1 (1966): 46.

22. As quoted in Rebecca Bell-Metereau, "Stealth, Sexuality, and Culture Status in *The Manchurian Candidate* and *Seconds*," in *A Little Solitaire: John Frankenheimer and American Film*, ed. Murray Pomerance and R. Barton Palmer (New Brunswick, NJ: Rutgers University Press, 2011), 57.

23. Frankenheimer was well aware of the resonance with McCarthyism; he specifically populated his film with blacklisted actors and crew members, including John Randolph, Jeff Carre, and Nathan Young. Ibid., 56.

24. James Quandt, *"The Face of Another:* Double Vision," Criterion Collection, July 9, 2007, https://www.criterion.com/current/posts/592-the-face-of-another-double-vision.

25. Ibid. Teshigahara and Abe's other collaborations are *Pitfall* (1962) and *Man without a Map* (1968).

26. For more on the Svengali trope, see Daniel Pick, *Svengali's Web: The Alien Enchanter in Modern Culture* (New Haven, CT: Yale University Press, 2000).

27. Bell-Metereau, "Stealth, Sexuality, and Culture Status," 55.

28. Haiken, *Venus Envy*, 15.

29. Lowenstein, *Shocking Representation*, 48.

30. Barbara Creed, "Horror and the Monstrous-Feminine: An Imaginary Abjection," *Screen* 27, no. 1 (1986): 44–71, doi:10.1093/screen/27.1.44.

31. Ethnographer David Sudnow coined the term "social death" in 1967 to refer to that state in which patients are treated essentially like corpses though they are still clinically or biologically alive. This was a year before the Harvard commission came up with the definition of "brain death," though in Sudnow's formulation they are quite similar. However, the term social death has broader purview and can be usefully applied to think about other definitions of life beyond the strictly and most basically biological. David Sudnow, *Passing On: The Social Organization of Dying* (Englewood Cliffs, NJ: Prentice-Hall, 1967), 74.

32. Giorgio Agamben, *Homo Sacer: Sovereign Power and Bare Life*, Meridian (Stanford, CA: Stanford University Press, 1998).

33. Arendt, *On Revolution*.

34. Appadurai, "Introduction: Commodities and the Politics of Value," 10. Organ donation is a gift that certainly entails obligation and reciprocity; recipients are well aware of their requirement to be worthy of that which they have been given. Appadurai is of course not the only anthropologist to explore this question, as I discuss in the previous chapter. For a discussion of gifting and organ transplants, see Lock, *Twice Dead*, 315. See also Matthew Sothern and Jen Dickinson, "Repaying the Gift of Life: Self-Help, Organ Transfer and the Debt of Care," *Social & Cultural Geography* 12, no. 8 (2011): 889–903, doi:10.1080/14649365.2011.624192.

35. Cornea, "Frankenheimer and the Science-Fiction/Horror Film," 232.

36. Identity transfer and the fracturing of the self remain an important concern for potential FAT patients. See for example Carosella and Pradeu, "Transplantation and Identity." I discuss this issue in greater depth in the previous chapter.

37. Alison Winter has discussed the shift from patient to experimental subject with respect to the O'Key sisters in Victorian England. See Alison Winter, *Mesmerized: Powers of Mind in Victorian Britain* (Chicago: University of Chicago Press, 1998), 99.

38. See for example Rhonda Gay Hartman, "Face Value: Challenges of Transplant Technology," *American Journal of Law & Medicine* 31, no. 1 (2005): 10–11.

39. The morality of makeup is neither obvious nor contiguous; as Annette Drew-Bear has shown, makeup was historically greeted with great suspicion precisely for what it was hiding. Drew-Bear, *Painted Faces on the Renaissance Stage: The Moral Significance of Face-Painting Conventions*. For more on the history of makeup, see Peiss, *Hope in a Jar*.

40. Lévinas and Nemo, *Ethics and Infinity*, 86–87.

41. Patrick McGrath, "Appearances to the Contrary: *Eyes without a Face*," Criterion Collection, July 20, 2014, http://www.criterion.com/current/posts/677-appearances-to-the-contrary-eyes-without-a-face.

42. Adam Lowenstein reads the feminization of violence in the film as an engagement with the Occupation and the Algerian War, both of which emasculated France and both of which occasioned many instances of violent emasculation. Lowenstein, *Shocking Representation*, 51.

43. McGrath, "Appearances to the Contrary."

44. Bell-Metereau, "Stealth, Sexuality, and Culture Status," 56.

45. Zombies, as liminal beings existing between life and death, past and present, are a rich site of scholarly inquiry. For more on the inbetweenness of zombies as related to difference, see Marc Leverette, "The Funk of Forty Thousand Years; Or, How the (Un)Dead Get Their Groove On," in *Zombie Culture: Autopsies of the Living Dead*, ed. Shawn McIntosh and Marc Leverette (Lanham, MD: Scarecrow Press, 2008), 185–212. Liminality and zombies have interesting applications to ethics and postcolonialism, as discussed in Robert Saunders, "Zombies in the Colonies: Imperialism and Contestation of Ethno-Political Space in Max Brooks's *The Zombie Survival Guide*," in *Monstrous Geographies: Places and Spaces of the Monstrous*, ed. Sarah Montin and Evelyn Tsitas (Oxford: Inter-Disciplinary Press, 2013), 19–46.

46. I refer specifically to sadomasochism, though there are other practices that fall into this category.

47. Journeys to nowhere are a strong organizing structure of the films of the 1960s. Think, for example, of *Psycho* (Alfred Hitchcock, 1960), in which Marion Crane's escape ends with her violent death. My profound thanks to Dana Polan for pointing me to this and other examples.

48. Once again, my gratitude to Dana Polan for this insightful reading.

49. Sennett, *The Fall of Public Man*.

50. For more on this narrative, see the previous chapter of this book.

51. We'll explore this balance in depth in chapter 5.

CHAPTER 4

1. The Cleveland Clinic first received IRB approval in 2004. The Centre Hospitalier Universitaire Nord in France also applied for ethics approval in 2004 but was denied permission though the rejection "left the door open for the nose and mouth 'triangle' to be transplanted," which was indeed what happened with the Isabelle Dinoire surgery in 2005. Rebecca Smith, "The World's First Face Transplant," *Evening Standard*, November 30, 2005, http://www.ucl.ac.uk/news/news-articles/inthenews/itn051202.

2. Peter Allen, "Face Transplant Woman Struggles with Identity," *Telegraph*, November 2, 2008, http://www.telegraph.co.uk/news/worldnews/europe/france/3367041/Face-transplant-woman-struggles-with-identity.html. For more details on the surgery itself, see "The First Face Transplant: Isabelle Dinoire," *My Multiple Sclerosis*, October 8, 2013, http://www.mymultiplesclerosis.co.uk/stranger-than-fiction/isabelledinoire.html.

3. This was no small lack; donors remain scarce across almost all transplants. Face-transplant donors present a particularly complicated set of matching criteria; in addition to blood type compatibility, there should to be some kind of gender, age, and skin-tone match, though these are in fact not medically necessary. Donor consent is also fraught, as the face is considered a special kind of donation that is not covered by standard organ-donation agreements.

4. Candidates for various organs are subject to different screening protocols, and most of

these differ from site to site. Face-transplant candidate screenings vary on the level of the particular hospital; a more universal screening protocol is one of the best-practice guidelines that are currently being discussed.

5. The closest analogue would be found in patients who undergo extreme plastic surgery, but these patients have a sense of what they will look like following the surgery. Some patients are inevitably disappointed in the outcome, and some do struggle to adjust to their new appearance, but these complications are different from those faced by facial-allograft recipients.

6. Seven of the most commonly cited concerns were: (1) rejection and drug-toxicity rates, (2) implications for donor population, (3) patient selection and compliance, (4) existence of other reconstructive options, (5) functional recovery, (6) psychological implications for the patient, and (7) informed consent. See Christian J. Vercler, "Ethical Issues in Face Transplantation," *Virtual Mentor* 12, no. 5 (2010): 378, doi:10.1001/virtualmentor.2010.12.5.jdsc1-1005.

7. For specific mortality rates in the United States, see "Dialysis and Transplant: Patient Survival Statistics, Quality Reports," *UW Health*, accessed June 23, 2015, http://www.uwhealth.org/transplant/dialysis-and-transplant-patient-survival-statistics/40583. American dialysis mortality rates are significantly higher than many other countries internationally. See Robert N. Foley and Raymond M. Hakim, "Why Is the Mortality of Dialysis Patients in the United States Much Higher than the Rest of the World?" *Journal of the American Society of Nephrology* 20, no. 7 (2009): 1432-35, doi:10.1681/ASN.2009030282.

8. For a close analysis of screening techniques in liver transplants, see Michael L. Volk et al., "Decision Making in Liver Transplant Selection Committees: A Multicenter Study," *Annals of Internal Medicine* 155, no. 8 (2011): 503-8, doi:10.7326/0003-4819-155-8-201110180-00006.

9. Samuel Taylor-Alexander, *On Face Transplantation: Life and Ethics in Experimental Biomedicine* (New York: Palgrave Macmillan, 2014).

10. Candidates for hand transplants face similar concerns; often parts of the arm have to be amputated to facilitate the transplant.

11. Clickbait refers to the digital practice of offering sensationalistic headlines without much detail that are designed to increase traffic or clicks with an eye to greater ad revenue. The KnowYourMeme website situates the first known use of the term on a blog post by Jay Geiger on December 1, 2006, in which he defines it as: "Any content or feature within a website that 'baits' a viewer to click. 'Anything interesting enough to catch a person's attention.' More often than not, clickbait uses 'highly alternative text/phrasing,' 'controversial slogans/ideas,' or 'culturally inspirational descriptions/events.' Clickbait is similar to linkbait but is generally seen as less effective, more shortsighted and more shortlived." See Don, "Clickbait," Know Your Meme, April 12, 2015, http://knowyourmeme.com/memes/clickbait. There is growing backlash against the practice; Facebook, for example, announced on August 14, 2014 that it had changed its News Feed algorithm to reduce clickbait in response to a study that reported 80 percent of readers "preferred headlines that helped them decide if they wanted to read the full article before they had to click through." See Khalid El-Arini and Joyce Tang, "News Feed FYI: Click-Baiting," Facebook Newsroom, August 25, 2014, http://newsroom.fb.com/news/2014/08/news-feed-fyi-click-baiting/.

12. Caplan coined the term in a lecture at the University of Western Ontario in 1987 about the harvesting of organs from anencephalic babies. There is strong resonance between the

"yuck (or yuk) factor" and Leon Kass's notion of "the wisdom of repugnance," though the two bioethicists have strongly diverging views on many topics, including embryo research. For more on this, see Matthew C. Nisbet, "The Competition for Worldviews: Values, Information, and Public Support for Stem Cell Research," *International Journal of Public Opinion Research* 17, no. 1 (2005): 90–112, doi:10.1093/ijpor/edh058.

13. Sara Ahmed, *Strange Encounters: Embodied Others in Post-Coloniality* (London: Routledge, 2013).

14. Françoise Baylis, "A Face Is Not Just Like a Hand: Pace Barker," *American Journal of Bioethics* 4, no. 3 (2004): 32, doi:10.1080/15265160490496804.

15. For a discussion of bioethical approaches to scarce resources in organ transplantation, see Arthur Caplan, "Organ Transplantation," in *The Hastings Center Bioethics Briefing Book for Journalists, Policymakers, and Campaigns*, ed. Mary Crowley (Garrison, NY: The Hastings Center, 2008), 129–32.

16. Kruvand, "Face to Face."

17. Joy M. Cypher, "Agency and the Dominant Face: Facial Transplantation and the Discourse of Normalcy," *American Communication Journal* 12 (Winter 2010), http://ac-journal.org/journal/pubs/2010/ACJproofCypher.pdf.

18. LaFrance, "'She Exists within Me'."

19. Taylor-Alexander, *On Face Transplantation*.

20. Taylor-Alexander also argues that the transplantation of the face as an organ challenges the special status of the face and equates it to all other transplantable organs. I disagree; the debates around FAT, as well as subsequent challenges that are unique to this transplant speak directly to the power of the face as a unique functional body part.

21. Edson C. Tandoc, "Journalism Is Twerking? How Web Analytics Is Changing the Process of Gatekeeping," *New Media & Society* 16, no. 4 (2014): 559–75, doi:10.1177/1461444814530541.

22. See for example a keynote address by the founder of BuzzFeed, Jonah Peretti, at the 2013 Disrupt NY conference that explains the financial advantages to clickbait: TechCrunch, "'Everyone Is Crazy' by Jonah Peretti (BuzzFeed) | Disrupt NY 2013 Keynote," 2013, https://www.youtube.com/watch?v=pPVpujeJwhY.

23. Gaye Tuchman, "Telling Stories," *Journal of Communication* 26, no. 4 (1976): 93–97, doi:10.1111/j.1460-2466.1976.tb01942.x.

24. Frank D. Durham, "News Frames as Social Narratives: TWA Flight 800," *Journal of Communication* 48, no. 4 (1998): 100–117, doi:10.1111/j.1460-2466.1998.tb02772.x.

25. This is not a quantitative or comprehensive study of all the news coverage of the event, though I did read all the English-language news through a LexisNexis search. Given the tremendous overlap in the stories both stylistically and in content, I was able to engage in close readings of selected texts that represent the broad trends. I focus on news reports rather than editorials, which represent a different set of journalistic and reporting decisions.

26. The Cleveland Clinic in the United States already had IRB approval for a face transplant, and Peter Butler's team at the Royal Free Hospital in the United Kingdom had come very close to approval at that time, though they did not ultimately achieve it. Australia had a different investment in the surgery: an Australian, Clint Hallam, was the first recipient of a hand transplant, and it was his surgeon who took the lead in the Dinoire case. I specifically did not draw

on the French newspaper coverage because their discussions incorporated an additional set of issues stemming from France being the first to successfully perform the surgery. The stakes for winning the face race are a central focus of much of the French coverage, making it a less-clear analogue than the other countries.

27. See for example Michael Mason, "A New Face: A Bold Surgeon, an Untried Surgery," *New York Times*, July 26, 2005, sec. Health.

28. Taylor-Alexander, *On Face Transplantation*; Kruvand, "Face to Face."

29. Kruvand, "Face to Face," 370.

30. Smith, "The World's First Face Transplant."

31. Jon Henley and Sarah Boseley, "Dog Attack Victim Gets World's First Face Transplant," *Guardian*, November 30, 2005, http://www.theguardian.com/science/2005/dec/01/france.medicineandhealth.

32. Lawrence K. Altman and Craig S. Smith, "French, in First, Use a Transplant to Repair a Face," *New York Times*, December 1, 2005, sec. Health.

33. Julie Robotham, "Facing up to a Superficial World Is No Vain Wish," *Sydney Morning Herald*, December 2, 2005, sec. Opinion, http://www.smh.com.au/news/opinion/facing-up-to-a-superficial-world-is-no-vain-wish/2005/12/01/1133422047549.html.

34. Lawrence K. Altman, "In a First, Doctors Transplant Part of a Face," *International Herald Tribune*, December 2, 2005.

35. Ibid.

36. "French Face Transplant Sparks Ethics Controversy," *St. Petersburg Times*, December 2, 2005, http://www.sptimes.com/2005/12/02/news_pf/Worldandnation/French_face_transplan.shtml.

37. Altman, "In a First, Doctors Transplant Part of a Face."

38. Henry Samuel and Nicole Martin, "Merci: Her First Word," *Telegraph*, December 3, 2005.

39. Daniel Billingham, "Woman's Thanks for Her New Face," *Birmingham Post*, December 3, 2005, http://www.thefreelibrary.com/Woman%27s+thanks+for+her+new+face.-a01393 90275.

40. Samuel and Martin, "Merci: Her First Word."

41. Ibid.

42. The quote was widely reported in a number of articles that same day, including those in the *Daily Star*, the *Scotsman*, the *Daily Mail*, and the *Times* (London). See Billingham, "Woman's Thanks for Her New Face"; Dan Newling and Peter Allen, "Face Swap Woman: The First Picture," *Daily Mail*, December 3, 2005, http://www.highbeam.com/doc/1G1-139384104.html; Adam Sage, "Woman with a New Face Looked in the Mirror and Said 'Merci,'" *Times* (London), December 3, 2005; Matthew Campbell, "'They Gave Me Back My Face,'" *Sunday Times* (London), December 4, 2005, http://www.thesundaytimes.co.uk/sto/news/uk_news/article208439.ece.

43. Campbell, "'They Gave Me Back My Face.'"

44. Newling and Allen, "Face Swap Woman: The First Picture"; Samuel and Martin, "Merci: Her First Word"; John Lichfield, "French Surgeons Defend Ethics of Face Transplant," *Independent*, December 3, 2005; Craig S. Smith, "From Transplant Patient with New

Face, 'Merci'; French Surgeons Pleased with Results," *International Herald Tribune*, December 3, 2005.

45. Newling and Allen, "Face Swap Woman: The First Picture."

46. Michael Mason and Lawrence K. Altman, "Ethical Concerns on Face Transplant Grow," *New York Times*, December 6, 2005, sec. International/Europe, http://www.nytimes.com/2005/12/06/international/europe/06facex.html.

47. Peter McKay, "Isabelle's New Face: Is It Moral?" *Daily Mail*, December 5, 2005, http://www.dailymail.co.uk/columnists/article-370623/Isabelles-new-face-Is-moral.html.

48. Ibid.

49. There are echoes here of vampires, parasites who drain vitality from others to maintain their immortality in a simulacra of life.

50. The discussions of the dangers of organ trafficking are far too numerous to list, but for a recent take, see Julie Bindel, "Organ Trafficking: A Deadly Trade," *Telegraph*, July 1, 2013, http://www.telegraph.co.uk/news/uknews/10146338/Organ-trafficking-a-deadly-trade.html.

51. For discussion of the various issues involved with a cash-for-organs system, see Sam Crowe, Eric Cohen, and Alan Rubenstein, "Increasing the Supply of Human Organs: Three Policy Proposals," The President's Council on Bioethics, February 2007, https://bioethicsarchive.georgetown.edu/pcbe/background/increasing_supply_of_human_organs.html.

52. There is also a gesture to the impossibility of informed consent from someone desperate, which I'll discuss later in the chapter.

53. John Cornwell, "The Face of Things to Come," *Sunday Times* (London), December 11, 2005, http://www.thesundaytimes.co.uk/sto/style/article158007.ece.

54. Kim Wilsher, "New Row Breaks out over Face Transplant," *Guardian*, December 5, 2005.

55. Lawrence K. Altman, "A Pioneering Transplant, and Now an Ethical Storm," *New York Times*, December 6, 2005.

56. Adam Sage, "Face Transplant Woman to Profit from Picture Sales," *Times* (London), December 8, 2005.

57. Around this time, newspapers published a number of editorials that echoed these various discussion and debates, often with a highly censorious tone. It took a few days for the press to get enough of a handle on the surgery to begin penning editorials, and these are interesting in their own right but do not reflect the framing issues that I focus on here. This analysis is restricted to the news reporting in the very first days following the surgery and how the framing itself shifted as more details emerged.

58. Despite the controversy, it is important to note longer-term perspectives: in 2014, the *Lancet* published an article examining the outcomes of all the face transplants to date, declaring the procedure itself a success overall. Saami Khalifian et al., "Facial Transplantation: The First 9 Years," *Lancet* 384, no. 9960 (2014): 2153–63, doi:10.1016/S0140-6736(13)62632-X.

59. Jon Bowen, "Gaining Face," *Salon*, May 19, 1999, http://www.salon.com/1999/05/19/face_transplants/.

60. For a timeline of all hand transplants to date, see "Hand Transplant History," *Composite Tissue Allotransplantation*, April 12, 2015, http://www.handtransplant.com/TheProcedure/HandTransplantHistory/tabid/96/Default.aspx.

61. Rachel Ellis, "First Hand Transplant Man Gives His Wife the Elbow . . . for His Nurse; Exclusive: Docs Sewed on New Limb . . . but He Hated It So Much He Had It Cut Off," *Sunday Mirror*, April 13, 2008, http://www.thefreelibrary.com/First+hand+transplant+man+gives+his+wife+the+elbow+..+for+his+nurse%3B . . .-a0177788748.

62. Andrew Buncombe, "I Hate My New Hand, Transplant Man Tells Doctors," *Independent*, October 20, 2000, http://www.independent.co.uk/life-style/health-and-families/health-news/i-hate-my-new-hand-transplant-man-tells-doctors-634222.html.

63. "Surgeons Amputate First Transplanted Hand," ABC News, January 6, 2006, http://abcnews.go.com/Health/story?id=117652&page=1.

64. François Petit et al., "Face Transplantation: Where Do We Stand?" *Plastic and Reconstructive Surgery* 113, no. 5 (2004): 1429–33.

65. I offer this not as a statement of blame and fault but as a reflection of the phenomenological experience of disability. There are those who engage in questions of best practice with respect to staring and disfigurement, for example Rosemarie Garland-Thomson, *Staring: How We Look* (Oxford: Oxford University Press, 2009). Here, I am interested in the ways in which the experience of facelessness and reactions to it frame the face transplant and help us understand the special status of the face. The challenges faced by the faceless are well documented online, in documentaries, and in a variety of memoirs. For a discussion of James Maki's life prior to his transplant, for example, see Susan Whitman Helfgot and William Novak, *The Match: Complete Strangers, a Miracle Face Transplant, Two Lives Transformed* (New York: Simon & Schuster, 2013).

66. For more on this indexical relationship, see Pearl, *About Faces*.

67. Henley and Boseley, "Dog Attack Victim Gets World's First Face Transplant."

68. Shehan Hettiaratchy and Peter E. M. Butler, "Face Transplantation—Fantasy or the Future?" *Lancet* 360, no. 9326 (2002): 5–6, doi:10.1016/S0140-6736(02)09361-3.

69. Laura Donnelly, "The High Price of Face Transplants," *Daily Telegraph*, March 26, 2011.

70. Ibid.

71. I discuss these issues in depth in chapter 2.

72. Hettiaratchy and Butler, "Face Transplantation—Fantasy or the Future?"

73. Much of what is at stake in these kinds of surgeries is the definition of non-necessity and the definition of risk. Doctors differ widely on these issues with respect to common interventions, such as elective caesarian sections, as well as riskier and experimental surgeries.

74. Maria Siemionow et al., "A Cadaver Study in Preparation for Facial Allograft Transplantation in Humans: Part I. What Are Alternative Sources for Total Facial Defect Coverage?" *Plastic and Reconstructive Surgery* 117, no. 3 (2006): 864–72; discussion 873–75, doi:10.1097/01.prs.0000204875.10333.56.

75. Kruvand, "Face to Face."

76. Siemionow et al., "A Cadaver Study."

77. Ibid., 870.

78. Ibid.

79. George J. Agich and Maria Siemionow, "Until They Have Faces: The Ethics of Facial Allograft Transplantation," *Journal of Medical Ethics* 31, no. 12 (2005): 707–9, doi:10.1136/jme.2005.011841.

80. Ibid., 707.
81. Ibid.
82. Ibid.
83. Ibid.
84. Baylis, "A Face Is Not Just Like a Hand," 32.
85. Agich and Siemionow, "Until They Have Faces," 708.
86. Ibid., 707. There are significant barriers to communication for people with facial paralysis and Moebius Syndrome, a condition in which people cannot make facial expressions. For more on these barriers, see Kathleen R. Bogart, Linda Tickle-Degnen, and Nalini Ambady, "Communicating without the Face: Holistic Perception of Emotions of People with Facial Paralysis," *Basic and Applied Social Psychology* 36, no. 4 (2014): 309–20, doi:10.1080/01973533.2014.917973.
87. Agich and Siemionow, "Until They Have Faces," 709.
88. Petit et al., "Face Transplantation," 1429.
89. Ibid.
90. Ibid., 1432.
91. Ibid.
92. I cover in depth the concurrent development of bioethics and organ transplantation in chapter 2.
93. Tia Powell, "Face Transplant: Real and Imagined Ethical Challenges," *Journal of Law, Medicine & Ethics* 34, no. 1 (2006): 111–15, doi:10.1111/j.1748-720X.2006.00014.x.
94. The 2003 report was superseded by the 2006 working report, which acknowledged the additional data from the two completed transplants but maintained a cautious view. It stipulated fifteen requirements that must be met for the facial transplantation to be sanctioned in a given institutional setting. The Royal College of Surgeons of England, Facial Transplantation Working Party Report, 2nd ed, November 2006, https://www.rcseng.ac.uk/publications/docs/facial_transplant_report_2006.html/@@download/pdffile/facial%20transplantation%202006.pdf.
95. The Royal College of Surgeons, "Facial Transplantation Working Party Report," November 2003, https://www.researchgate.net/publication/7053799_Facial_transplantation_A_working_party_report_from_the_Royal_College_of_Surgeons_of_England
96. David Magnus et al., "A New Era for AJOB," *American Journal of Bioethics* 4, no. 3 (2004): x–xi.
97. Hartman, "Face Value."
98. Ibid., 7.
99. Ibid., 14.
100. Ibid.
101. Ibid., 14–15.
102. Ibid., 15. The question of using desperate patients for pioneering research and clinical trials is well rehearsed in the bioethical literature. See Peter Allmark and Su Mason, "Should Desperate Volunteers Be Included in Randomised Controlled Trials?" *Journal of Medical Ethics* 32, no. 9 (2006): 548–53, doi:10.1136/jme.2005.014282; Charles Bosk, "Obtaining Voluntary Consent for Research in Desperately Ill Patients," *Medical Care* 40, no. 9 (2002): 64–68; Marcela G. del Carmen and Steven Joffe, "Informed Consent for Medical Treatment and Research:

A Review," *Oncologist* 10, no. 8 (2005): 636–41, doi:10.1634/theoncologist.10-8-636; Mary M. Flannery, "Research on the Terminally Ill: A Balancing Act between Facilitating Access to Innovative Therapies and Protecting Vulnerable Subjects in Search of One Last Hope for Survival," Harvard University, 2003, http://nrs.harvard.edu/urn-3:HUL.InstRepos:8852203; Teresa Swift, "Desperation May Affect Autonomy but Not Informed Consent," *American Journal of Bioethics Neuroscience* 2, no. 1 (2011).

103. Luis Bermudez, "Face Transplant: Is It Worth It?" *Plastic and Reconstructive Surgery* 117, no. 6 (2006): 1891, doi:10.1097/01.prs.0000219070.92470.0e.

104. Ibid., 1892–93.

105. Ibid., 1893.

106. This issue is well highlighted in the Royal College of Surgeons' two working reports on facial transplants.

107. Bermudez, "Face Transplant," 1895.

108. John H. Barker et al., "Investigation of Risk Acceptance in Facial Transplantation," *Plastic and Reconstructive Surgery* 118, no. 3 (2006): 663–70, doi:10.1097/01.prs.0000233202.98336.8c.

109. See for example Mary B. Prendergast and Robert S. Gaston, "Optimizing Medication Adherence: An Ongoing Opportunity to Improve Outcomes after Kidney Transplantation," *Clinical Journal of the American Society of Nephrology: CJASN* 5, no. 7 (2010): 1305–11, doi:10.2215/CJN.07241009.

110. Jon W. Jones et al., "Successful Hand Transplantation—One-Year Follow-Up," *New England Journal of Medicine* 343, no. 7 (2000): 468–73, doi:10.1056/NEJM200008173430704; G. R. Tobin et al., "Ethical Considerations in the Early Composite Tissue Allograft Experience: A Review of the Louisville Ethics Program," *Transplantation Proceedings* 37, no. 2 (2005): 1392–95, doi:10.1016/j.transproceed.2004.12.179.

111. The next chapter explores the post-face-transplant experience in depth.

112. See chapter 2.

113. Benoît Lengelé et al., "Facing Up Is an Act of Dignity: Lessons in Elegance Addressed to the Polemicists of the First Human Face Transplant," *Plastic and Reconstructive Surgery* 120, no. 3 (2007): 803–6, doi:10.1097/01.prs.0000271097.22789.79.

114. Samuel and Martin, "Merci: Her First Word."

115. Kruvand, "Face to Face," 373.

116. Ibid.

117. "Donor for World's First Face Transplant Had Hanged Herself," *Independent*, accessed July 15, 2015, http://www.independent.co.uk/news/world/europe/donor-for-worlds-first-face-transplant-had-hanged-herself-518079.html.

118. Lengelé et al., "Facing Up Is an Act of Dignity," 802.

CHAPTER 5

1. Graeme Turner, *Understanding Celebrity* (Thousand Oaks, CA: Sage Publications, 2013), 6.

2. Cressida J. Heyes, *Self-Transformations: Foucault, Ethics, and Normalized Bodies* (Oxford: Oxford University Press, 2007), 96–105.

3. See for example Heyes, *Self-Transformations*; Sender, *The Makeover*.

4. Koswara died on January 30, 2016 as a result of his condition. "Half Man Half Tree," *Extraordinary People*, Season 6, Episode 15, YouTube video, from a documentary televised by TLC on April 14, 2008, posted by DadoTheGoodVillian, September 29, 2015, https://www.youtube.com/watch?v=GIKe4aFzln0; Ben Shackleford, "Separating Twins," *NOVA*, PBS, February 8, 2012, http://www.pbs.org/wgbh/nova/body/separating-twins.html.

5. Tania Lewis, "Revealing the Makeover Show," *Continuum* 22, no. 4 (2008): 442, doi:10.1080/10304310802190053.

6. *Queer Eye for the Straight Guy* stands as an example of a show that specifically targets men for personal and lifestyle makeovers. As Katherine Sender points out, this program does play with normative categories but ultimately reinforces traditional heteronormative and stereotypical performances of gender. Sender, *The Makeover*, 34.

7. Shows like *The Biggest Loser* require extreme physical and psychological feats, much of which is obscured in the televised portion. Extreme plastic surgery shows like *The Swan* and *Extreme Makeover* are also highly physically demanding. Many of the less explicitly physical shows also require a great deal of energy, both of the mind and the body.

8. These details are explored in depth in Sender, *The Makeover*.

9. Meredith Jones, "Media-Bodies and Screen-Births: Cosmetic Surgery Reality Television," *Continuum* 22, no. 4 (2008): 515–24, doi:10.1080/10304310802189998.

10. Wegenstein and Ruck, "Physiognomy, Reality Television and the Cosmetic Gaze."

11. There have been very few cases of rejection to date, and many of them were dealt with by massive doses of antirejection medication, including a recent episode involving Carmen Tarleton. See Liz Kowalczyk, "Transplant Recipient Came Close to Losing Donor Face to Rejection," *Boston Globe*, May 1, 2013, https://www.bostonglobe.com/lifestyle/health-wellness/2013/04/30/transplant-recipient-came-close-losing-donor-face-rejection/nSLWrzivtDqitAn3cKevAL/story.html. For more information on medical rejection, see Hettiaratchy and Butler, "Face Transplantation—Fantasy or the Future?"

12. A *GQ* article on Richard Lee Norris was written in very much this kind of moralistic tone, judging Norris for his post-transplant bodily transgressions. Jeanne Marie Laskas, "The Miraculous Face Transplant of Richard Norris," *GQ*, August 28, 2014, http://www.gq.com/news-politics/newsmakers/201408/richard-norris.

13. Ted Kallmyer, "Biggest Loser Then and Now: Have Former Winners Kept the Weight Off?" *Healthy Eater*, accessed April 28, 2015, http://healthyeater.com/biggest-loser-then-now.

14. Ted Kallmyer, "Former Contestant Kai Hibbard Slams *The Biggest Loser*," *Healthy Eater*, accessed April 28, 2015, http://healthyeater.com/hibbard-slams-biggest-loser.

15. Sean Daly People, "Former '*Swan*' Contestant Is Now a 300 Pound Shut In," TheTVPage.com, April 28, 2015, http://thetvpage.com/2013/02/21/former-swan-contestant-is-now-a-300-pound-shut-in/.

16. Reality-television participants also try to parlay their exposure into more lasting platforms, either as advocates for related issues (like fitness and weight loss), for marketing and commercial projects, or in the entertainment industry generally.

17. van Dijck, *The Transparent Body*.

18. David L. Clark and Catherine Myser, "'Fixing' Katie and Eilish: Medical Documenta-

ries and the Subjection of Conjoined Twins," *Literature and Medicine* 17, no. 1 (1998): 45–67, doi:10.1353/lm.1998.0004.

19. Ibid.; David Clark and Catherine Myser, "Being Humaned: Medical Documentaries and the Hyperrealization of Conjoined Twins," in *Freakery: Cultural Spectacles of the Extraordinary Body*, ed. Rosemarie Garland Thomson (New York: New York University Press, 1994), 338–55; van Dijck, *The Transparent Body*.

20. Shildrick, "Corporeal Cuts."

21. Clark and Myser, "Being Humaned," 339.

22. Of course, operations don't always make things better. Sometimes one or both of the conjoined twins die as a result of the operation, which always render the patients' bodies more vulnerable.

23. Clark and Myser, "Being Humaned," 350.

24. Scott McQuire, "From Glass Architecture to Big Brother: Scenes from a Cultural History of Transparency," *Cultural Studies Review* 9, no. 1 (2003): 116.

25. Sender, *The Makeover*, 29.

26. "Brigham Face Transplant Recipient Goes Home," *Harvard Gazette*, accessed June 24, 2015, http://news.harvard.edu/gazette/story/2009/05/brigham-face-transplant-recipient-goes-home/.

27. Diane Sawyer, "Connie Culp Talks About Her Future," *Good Morning America*, ABC News, May 8, 2009. accessed March 13, 2015, https://www.youtube.com/watch?v=17nD4FowE10&feature=youtube_gdata_player.

28. Marjorie Kruvand explores some of the framing reasons for the difference in news coverage between Maki and Dinoire. Kruvand, "Face to Face."

29. Liz Kowalczyk, "His Tragic Accident behind Him, New England's First Face Transplant Patient Tells of an Arduous Journey and a Life Renewed," *Boston Globe*, May 21, 2009, http://www.boston.com/news/local/massachusetts/articles/2009/05/21/his_tragic_accident_behind_him_new_englands_first_face_transplant_patient_tells_of_an_arduous_journey_and_a_life_renewed/.

30. Emma North, "Face Transplant Man's Plea," *BBC News* video, May 23, 2009, accessed March 13, 2015, http://news.bbc.co.uk/2/hi/americas/8064820.stm.

31. Diane Sawyer, "Jim Maki: New Life With a New Face," *Good Morning America*, ABC News, June 18, 2009, accessed March 4, 2013, https://www.youtube.com/watch?v=s4D_D2pPFy0&feature=youtube_gdata_player.

32. Maki's smooth integration of his new face stands in sharp contrast to Isabelle Dinoire's struggles to come to terms with bearing and continually seeing the face of another. Although not a face-transplant recipient, poet and memoirist Lucy Grealy's story of cancer of the jaw and subsequent reconstructions also highlights the challenges of coming to terms with a new face. Grealy's narrative emphasizes that every attempted gain was also a loss: as doctors tried to reconstruct her disfigured face, they mutilated her body, cutting and grafting from one place to fill in another. Lucy Grealy and Ann Patchett, *Autobiography of a Face* (New York: Harper Perennial, 2003).

33. Sawyer, "Connie Culp Talks About Her Future."

34. Diane Sawyer, "The New Face of the First American Face Transplant," ABC News,

August 26, 2010, accessed March 4, 2015, http://abcnews.go.com/WNT/video/face-american-face-transplant-connie-culp-surgery-recovery-surgery-diane-sawyer-11491441.

35. Diane Sawyer, "Connie Culp Meets Donor's Family," *Good Morning America*, ABC News, December 21, 2010, accessed March 4, 2015, http://abcnews.go.com/GMA/video/face-transplant-recipient-connie-culp-meets-donor-family-12448484?page=1.

36. JuJu Chang, "Dallas Wiens Smiling 10 Months after Full Face Transplant," *Good Morning America*, ABC News, January 15, 2012, accessed March 4, 2015, http://abcnews.go.com/Health/Wellness/dallas-wiens-smiling-full-face-transplant/story?id=15366582.

37. Weins's blindness would have excluded him from candidacy for the surgery based on the Cleveland Clinic's screening criteria; he had his procedure at Brigham and Women's Hospital. To date, there are no universal protocols or evaluative procedures for selecting candidates for the surgery, nor are there agreed-upon best-practice guidelines. Chad R. Gordon et al., "The Cleveland Clinic FACES Score: A Preliminary Assessment Tool for Identifying the Optimal Face Transplant Candidate," *The Journal of Craniofacial Surgery* 20, no. 6 (2009): 1969–74, doi:10.1097/SCS.0b013e3181bd2c86.

38. "Soldier Who Suffered Horrific Face Injury when He Was Electrocuted after Saving Woman Receives Face Transplant," *Mail Online*, accessed April 30, 2015, http://www.dailymail.co.uk/news/article-1380851/Soldier-Mitch-Hunter-electrocuted-saving-woman-gets-face-transplant.html.

39. Michael Mosley, "Meeting Face Transplant Patient, Mitch Hunter," *Frontline Medicine: Rebuilding Lives*, BBC Two, November 27, 2011, accessed March 4, 2015, http://www.bbc.com/news/health-15871905.

40. Mitch Hunter, "Mitch Hunter Face Transplant Q&A Part 2," YouTube video, 8:13, June 19, 2014, https://www.youtube.com/watch?v=g6ag5zB4RIY.

41. Jessica Hayes, "Mitch Hunter Returns to Indy," 10 pm News, WISH-TV June 5, 2011, https://www.youtube.com/watch?v=Ud-W9QuQ12U&feature=youtube_gdata_player. Hunter has received more press recently as he offered various updates on his condition five years later.

42. "Mitch Hunter (Full Face Transplant) • /r/IAmA," Reddit, accessed May 14, 2015, http://www.reddit.com/r/IAmA/comments/1e4023/mitch_hunter_full_face_transplant/.

43. Hunter, "Mitch Hunter Face Transplant Q&A Part 2"; "Mitch Hunter (Full Face Transplant) • /r/IAmA."

44. Sarah Wallace, "Charla Nash: New Face of Chimpanzee Attack Victim Revealed," *Eyewitness News*, WABC, August 11, 2011, http://abcnews.go.com/Health/Wellness/charla-nash-face-chimpanzee-attack-victim-revealed/story?id=14282402.

45. "The Will to Live," *Oprah*, accessed May 5, 2015, http://www.oprah.com/oprahshow/Chimp-Attack-Victim-Charla-Nash-Shows-Her-Face.

46. "Charla Nash, Victim of Chimpanzee Attack, Speaks Out for Primate Safety," YouTube video, 7:45, posted by "The Humane Society," July 11, 2014, accessed May 5, 2015, https://www.youtube.com/watch?v=2O464ys9E60; "Charla Nash, An American Girl," YouTube video, 6:24, posted by Shelly Sindland Media, March 9, 2015, accessed May 5, 2015, https://www.youtube.com/watch?v=cYDt7hWzF_Y.

47. Laskas, "The Miraculous Face Transplant of Richard Norris."

48. Ann Curry, "Ann Curry Reports: A Face in the Crowd, Part 1," NBC News, June 28, 2013, http://www.nbcnews.com/video/ann-curry-reports/52347670.

49. Laskas, "The Miraculous Face Transplant of Richard Norris."

50. Jessica Tozer, "Changing the Face of Modern Medicine," Armed with Science, January 9, 2014, http://science.dodlive.mil/2014/01/09/changing-the-face-of-modern-medicine/.

51. Karen Lancaster, "University of Maryland Patient Exceeding Expectations after Face Transplant," University of Maryland Medical Center, October 16, 2012, http://umm.edu/news-and-events/news-releases/2012/um-patient-exceeding-expectations-after-face-transplant.

52. Katie Drummond, "Beyond Recognition: The Incredible Story of a Face Transplant," *Verge*, June 4, 2013, http://www.theverge.com/2013/6/4/4381890/carmen-tarleton-the-incredible-story-of-a-face-transplant.

53. "Carmen's Survival Story," *The Doctors*, CBS, accessed May 6, 2015, www.thedoctorstv.com/videos/carmen-s-survival-story.

54. Jeremy Schaap, "Carmen: A Survivors Story," *E:60*, ESPN, October 11, 2014, http://espn.go.com/video/clip?id=11854612.

55. Ibid.

56. Sender, *The Makeover*, 27.

57. Stephen Kiehl, "Her New Face Gives Woman a 'Normal Life,'" *Baltimore Sun*, February 7, 2006, http://articles.baltimoresun.com/2006-02-07/features/0602070074_1_facial-transplant-face-forward-moss.

58. Allen, "Face Transplant Woman Struggles with Identity." Dinoire's resistance to her new face is covered in depth in LaFrance, "'She Exists within Me.'"

59. Barbara Davis and Peter Allen, "I Still Fear My Own Reflection, Says the World's First Face Transplant Patient," *Daily Mail Online*, October 5, 2007, http://www.dailymail.co.uk/femail/article-485990/I-fear-reflection-says-worlds-face-transplant-patient.html.

60. Kiehl, "Her New Face Gives Woman a 'Normal Life.'"

61. Ralf Smeets et al., "Face Transplantation: On the Verge of Becoming Clinical Routine?" *BioMed Research International* (2014): e907272, doi:10.1155/2014/907272.

62. As I discuss below, there are a number of matching criteria that have to be met between donor and recipient. And while, technically, those who have consented to organ donation have also consented to donate their faces, to date hospitals have agreed that faces are a special case that require explicit additional familial consent.

63. Aetna, *Clinical Policy Bulletin: Face Transplantation Number: 0819*, March 13, 2015. The policy will be reviewed again in October 2015.

64. There are a number of matching criteria that have to be met for a donor to be suitable for a given potential recipient, including blood type, age, gender, and skin color. These last three categories are not medically vital; age does not have to be an exact match but rather must fall within a reasonable range. Many surgeons have claimed that cross-gender transplants could in fact be highly successful given that the donor face adheres to the recipient's bone structure, but they remain concerned about the psychological implication for such a procedure. Skin color is likewise a question of cultural and aesthetic norms that could present significant psychological challenges to a non-matching recipient. Recipients have little information about their donors prior to their surgeries beyond these basic criteria and would not have enough information to

veto a potential match on the basis of aesthetics, should such a veto even be an option beyond hesitation around undergoing the surgery itself. Chad R. Gordon et al., "Overcoming Cross-Gender Differences and Challenges in Le Fort-Based, Craniomaxillofacial Transplantation with Enhanced Computer-Assisted Technology," *Annals of Plastic Surgery* 71, no. 4 (2013): 421–28, doi:10.1097/SAP.0b013e3182a0df45.

65. Pushkal P. Garg et al., "Impact of Gender on Access to the Renal Transplant Waiting List for Pediatric and Adult Patients," *Journal of the American Society of Nephrology* 11, no. 5 (2000): 958–64; Lynne Young and Maureen Little, "Women and Heart Transplantation: An Issue of Gender Equity?" *Health Care for Women International* 25, no. 5 (2004): 436–53, doi:10.1080/07399330490272778.

66. Sender, *The Makeover*, 27.

67. To date, only one US face-transplant recipient has remained anonymous.

68. Tozer, "Changing the Face of Modern Medicine."

69. Marvin, *Blood Sacrifice and the Nation*.

70. Agamben, *Homo Sacer: Sovereign Power and Bare Life*.

71. Tozer, "Changing the Face of Modern Medicine."

72. Rachel Feltman, "The Military Has High Hopes for Face Transplants," *Washington Post*, March 10, 2015, http://www.washingtonpost.com/news/speaking-of-science/wp/2015/03/10/the-military-has-high-hopes-for-face-transplants/.

73. Lancaster, "University of Maryland Patient Exceeding Expectations after Face Transplant."

74. We have echoes, here, of World War I attempts to treat soldiers for shell shock and neurasthenia in order to return them to the front.

75. Tozer, "Changing the Face of Modern Medicine."

76. Ibid.

77. Tania Lewis, *Smart Living: Lifestyle Media and Popular Expertise* (New York: Peter Lang, 2008).

78. Jones, "Media-Bodies and Screen-Births."

CHAPTER 6

1. I refer here to the "better than well" debates sparked by the widespread prescription of antidepressant medication like Prozac for users who were not clinically depressed but found that they preferred themselves while medicated, operating in a state called better than well. Carl Elliott and Peter D. Kramer, *Better Than Well: American Medicine Meets the American Dream*, rep. ed. (New York: WW Norton, 2004).

2. All of these questions are worthy of (and have garnered) books of their own, too numerous to list. For one overview, see David Rose, *Enchanted Objects: Design, Human Desire, and the Internet of Things* (New York: Simon and Schuster, 2014).

3. For a compelling critique of so-called colorblindness with respect to race, see Eduardo Bonilla-Silva, *Racism without Racists: Color-Blind Racism and the Persistence of Racial Inequality in the United States* (Lanham, MD: Rowman & Littlefield, 2006).

4. Emmanuel Lévinas, *Ethics and Infinity: Conversations with Philippe Nemo*, trans. Richard A. Cohen (Pittsburgh: Duquesne University Press, 1985), 354, 85.

5. Amit Pinchevski, "Lévinas as a Media Theorist: Toward an Ethics of Mediation," *Philosophy and Rhetoric* 47, no. 1 (2014): 48–72.

6. Deleuze and Guattari, *A Thousand Plateaus*.

7. Margrit Shildrick, "'Why Should Our Bodies End at the Skin?': Embodiment, Boundaries, and Somatechnics," *Hypatia* 30, no. 1 (2015): 13–29, doi:10.1111/hypa.12114.

8. Deleuze and Guattari, *A Thousand Plateaus*.

9. For this phrasing, I remain deeply indebted to Sun-Ha Hong.

10. Emmanuel Lévinas, *Totality and Infinity: An Essay on Exteriority* (The Hague: Martinus Nijhoff, 1979).

11. Deleuze and Guattari, *A Thousand Plateaus*; Jürgen Habermas, *The Structural Transformation of the Public Sphere: An Inquiry into a Category of Bourgeois Society*, rep. ed. (Cambridge, MA: MIT Press, 1991); Butler, *Precarious Life*.

12. Harvey "Two-Face" Dent is a supervillain and Batman former ally who becomes a criminal and nemesis after an acid attack from a mob boss that disfigures the left side of his face. He flips a coin to decide whether to enact his good side or evil side when making decisions. Two-Face was first introduced in August 1942.

13. Bordo, *Unbearable Weight*.

14. Sometimes high-tech prostheses are less effective than simpler models. For a discussion of the challenges of more advanced prostheses, see Rose Eveleth, "When State-of-the-Art Is Second Best," *NOVA Next*, PBS, March 5, 2014, http://www.pbs.org/wgbh/nova/next/tech/durable-prostheses/.

15. Sherry Turkle, *Life on the Screen* (New York: Simon and Schuster, 2011); Howard Rheingold, *The Virtual Community: Homesteading on the Electronic Frontier* (Cambridge, MA: MIT Press, 1993). For a critique of Turkle, see Lori Kendall, *Hanging Out in the Virtual Pub: Masculinities and Relationships Online* (Berkeley: University of California Press, 2002). More recent performance-based analyses include Brenda Laurel, *Computers as Theatre* (Upper Saddle River, NJ: Pearson Education, 2013).

16. Lisa Nakamura, *Cybertypes: Race, Ethnicity, and Identity on the Internet* (New York: Routledge, 2002).

17. Yoel Roth, "'No Overly Suggestive Photos of Any Kind': Content Management and the Policing of Self in Gay Digital Communities," *Communication, Culture & Critique*, 8, no. 3 (2015): 414–32, doi:10.1111/cccr.12096.

18. Lauren F. Sessions, "'You Looked Better on MySpace': Deception and Authenticity on the Web 2.0," *First Monday* 14, no. 7 (2009), doi:10.5210/fm.v14i7.2539.

19. Allucquere Rosanne Stone, "Will the Real Body Please Stand Up? Boundary Stories about Virtual Cultures," in *Cyberspace: First Steps*, ed. Michael L. Benedikt (Cambridge, MA: MIT Press, 1991), 81–118.

20. Some of the cases of identity play are of course quite playful with few high-stakes consequences, but there are far more dangerous examples. Catfishing is only the tip of the iceberg; there are numerous cases of pedophiles using the Internet to lure young people into danger, in addition to multiple other dangerous scenarios. While these interactions are the exception, they are chilling and worth noting.

21. As Amit Pinchevski and John Durham Peters have more recently shown, the digital space can be transformative for people on the autism spectrum. Amit Pinchevski and John

Durham Peters, "Autism and New Media: Disability between Technology and Society," *New Media & Society* (2015), 1–17, doi:10.1177/1461444815594441.

22. Jessa Lingel and Tarleton Gillespie, "One Name to Rule Them All: Facebook's Identity Problem," *Atlantic*, October 2, 2014, http://www.theatlantic.com/technology/archive/2014/10/one-name-to-rule-them-all-facebook-still-insists-on-a-single-identity/381039/.

23. Rena Bivens, "The Gender Binary Will Not Be Deprogrammed: Ten Years of Coding Gender on Facebook," (SSRN scholarly paper, Social Science Research Network, Rochester, NY, April 30, 2014), http://papers.ssrn.com/abstract=2431443.

24. Sennett, *The Fall of Public Man*.

25. Goffman, *The Presentation of Self*.

26. Barry Smart, "Facing the Body—Goffman, Lévinas and the Subject of Ethics," *Body & Society* 2, no. 2 (1996): 67–78, doi:10.1177/1357034X96002002004.

27. I refer here to the condition of being faceless IRL, though, as the tradition of headless/torso pictures on the dating site Grindr shows, one can also be quite literally faceless online.

28. Amit Pinchevski, "Ethics on the Line," *Southern Communication Journal* 68, no. 2 (2003): 152–66, doi:10.1080/10417940309373257.

29. Roger Silverstone, *Media and Morality: On the Rise of the Mediapolis* (Cambridge: Polity, 2006); Lucas D. Introna, "The Face and the Interface: Thinking with Lévinas on Ethics and Justice in an Electronically Mediated World" (working paper, Department of Organisation, Work, and Technology, Lancaster University, Lancaster, 2003).

30. John Durham Peters, *Speaking into the Air: A History of the Idea of Communication* (Chicago: University of Chicago Press, 2001).

31. Peters writes: "If success in communication was once the art of reaching across the intervening bodies to touch another's spirit, in the age of electronic media it has become the art of reaching across the intervening spirits to touch another body. Not the ghost in the machine, but the body in the medium is the central dilemma of modern communications." Ibid., 224–25.

32. See chapter 5 for more on Weins.

33. The singer/songwriter Sia plays with the publicity of the face by covering hers with her wig. She has explained that she hides her face in public as a way to maintain privacy and avoid celebrity; one can read in her decision an attempt to evade a loss of power over her own image and own identity. Her face, like all faces, is always public; she chooses not to show it for that very reason, though she has also become more famous for that very reason. Lavanya Ramanathan, "Sia Is Rejecting Fame. Which Has Made Her More Famous than Ever," *Washington Post*, February 6, 2015, http://www.washingtonpost.com/lifestyle/style/sia-is-rejecting-fame-which-has-made-her-more-famous-than-ever/2015/02/06/419665da-aafc-11e4-abe8-e1ef60ca26de_story.html. There are other ways of hiding the face as well, including religious practices of veiling and the use of masks; these do not make the face any less public. They make the wearer publicly faceless.

34. Julian Dibbell, *My Tiny Life: Crime and Passion in a Virtual World* (New York: Holt Paperbacks, 1999).

35. Pearl, *About Faces*.

36. Haraway, "A Cyborg Manifesto," 220.

37. Marc Shell, "Moses' Tongue," *Common Knowledge* 12, no. 1 (2006): 150–76.

38. Exod. 33:19–23, Hebrew Bible.
39. For more on touch as communication, see Peters, *Speaking into the Air*.
40. Martin Buber, *I and Thou* (Mansfield Centre, CT: Martino Fine Books, 2010).
41. Emmanuel Lévinas, "The Trace of the Other," in *Deconstruction in Context: Literature and Philosophy*, ed. Mark C. Taylor (Chicago: University of Chicago Press, 1986), 345–59; Gen. 1:27, Hebrew Bible.
42. G. W. F. Hegel, *Phenomenology of Spirit*, trans. A. V. Miller (Oxford: Oxford University Press, 1977).
43. Pinchevski, "Levinas as a Media Theorist."
44. Exod. 3:4, Hebrew Bible.
45. The medieval commentator Rashi (Rabbi Shimon Yitzchaki) interprets Exod. 19:17 to say that the Israelites were standing beneath the mountain when they received the Torah. He bases his commentary on Talmud Bavli Shabbat 88a.
46. Deleuze and Guattari, *A Thousand Plateaus*.
47. Garland-Thomson, *Staring: How We Look* (Oxford: Oxford University Press, 2009).
48. Bordo, *Unbearable Weight*.
49. Haraway, "A Cyborg Manifesto," 181.
50. Bordo, *Unbearable Weight*; Shildrick, "Corporeal Cuts"; Garland-Thomson, "Integrating Disability, Transforming Feminist Theory."
51. Perry, "Buying White Beauty."
52. Davis, *Reshaping the Female Body*.
53. *Kingsman: The Secret Service* directed by Matthew Vaughn (Los Angeles: 20th Century Fox, 2015), DVD.
54. Jon Favreau, dir. *Iron Man* (2008).
55. Roslyn Sulcas, "Matthew Vaughn and Jane Goldman Discuss 'Kingsman,'" *New York Times*, February 6, 2015, http://www.nytimes.com/2015/02/08/movies/matthew-vaughn-and-jane-goldman-discuss-kingsman.html.
56. Film has a long history of manipulating difference; most recently, Emma Stone was cast as a woman with native Hawai'ian ancestry in the film *Aloha*, causing a great deal of controversy.
57. Too numerous to list in full, and too tokenistic to summarize.
58. Bordo, *Unbearable Weight*.
59. Jones, "Media-Bodies and Screen-Births."
60. Warwick Mules, "This Face: A Critique of Faciality as Mediated Self-Presence," *Transformations*, no. 18 (2010).
61. See for example Robert Ayers, "Serene and Happy and Distant: An Interview with Orlan," *Body & Society* 5, nos. 2–3 (1999): 171–84, doi:10.1177/1357034X99005002010; Alyda Faber, "Saint Orlan: Ritual as Violent Spectacle and Cultural Criticism," *TDR/The Drama Review* 46, no. 1 (2002): 85–92, doi:10.1162/105420402753555868; Genesis P-Orridge, *Painful but Fabulous: The Life and Art of Genesis P-Orridge* (New York: Soft Skull Press, 2002); Anja Zimmermann, "'Sorry for Having to Make You Suffer': Body, Spectator, and the Gaze in the Performances of Yves Klein, Gina Pane, and Orlan," *Discourse* 24, no. 3 (2002): 27–46, doi:10.1353/dis.2003.0035; Emma Govan, "Visceral Excess: Cosmimesis in the Work of Orlan," *Women & Performance* 15, no. 2 (2005): 147–59, doi:10.1080/07407700508571509; C. Jill

O'Bryan, *Carnal Art: Orlan's Refacing* (Minneapolis: University of Minnesota Press, 2005); Krista Miranda, "DNA, AND: A Meditation on Pandrogeny," *Women & Performance* 20, no. 3 (2010): 347-53, doi:10.1080/0740770X.2010.529266.

62. Gianna Bouchard, "Incisive Acts: Orlan Anatomised," in *Orlan: A Hybrid Body of Artworks*, ed. Simon Donger, Simon Shepherd, and Orlan (New York: Routledge, 2010), 62-73.

63. Panayiota Chrysochou, "Orlan's Gruesome Theatre: Religion, Technology and the Politics of Female Self-Mutilation and Violence as Gestus," in *Female Beauty in Art: History, Feminism, Women Artists*, ed. Maria Ioannou and Maria Kyriakidou (Newcastle upon Tyne: Cambridge Scholars, 2014), 67-98.

64. Imogen Ashby, "The Mutant Woman: The Use and Abuse of the Female Body in Performance Art," *Contemporary Theatre Review* 10, no. 3 (2000): 39-51, doi:10.1080/10486800008568595.

65. P-Orridge, *Painful but Fabulous: The Life and Art of Genesis P-Orridge*, 49.

66. Chrysochou, "Orlan's Gruesome Theatre."

67. Joanna Zylinska and Gary Hall, "Probings: An Interview with Stelarc," in *The Cyborg Experiments: The Extensions of the Body in the Media Age*, ed. Joanna Zylinska (New York: Bloomsbury Academic, 2002), 114-30.

68. Jones, *Skintight*; Kobena Mercer, "Monster Metaphors: Notes on Michael Jackson's 'Thriller,'" *Screen* 27, no. 1 (1986): 26-43, doi:10.1093/screen/27.1.26; Kathy Davis, "Surgical Passing: Or Why Michael Jackson's Nose Makes 'Us' Uneasy," *Feminist Theory* 4, no. 1 (2003): 73-92; Roberta Mock, "Stand-up Comedy and the Legacy of the Mature Vagina," *Women & Performance: A Journal of Feminist Theory* 22, no. 1 (2012): 9-28, doi:10.1080/0740770X.2012.685394; Louise Peacock, "Joan Rivers—Reading the Meaning," *Comedy Studies* 2, no. 2 (2011): 125-37, doi:10.1386/cost.2.2.125_1; Martha Gever, "The Trouble with Moralism: Nip/Tuck," *Media, Culture & Society* 32, no. 1 (2010): 105-22; Bernadette Wegenstein, *The Cosmetic Gaze: Body Modification and the Construction of Beauty* (Cambridge, MA: MIT Press, 2012).

69. "Joan Rivers," *Nip/Tuck*, directed by Ryan Murphy, (FX, October 5, 2004), DVD.

70. For a list of some of Rivers's best-known plastic surgery jokes, see *THR* staff, "Joan Rivers Dead: Best Quotes about Plastic Surgery," *The Hollywood Reporter*, September 4, 2014, http://www.hollywoodreporter.com/gallery/joan-rivers-dead-best-quotes-730303.

71. Davis, "Surgical Passing."

72. Mercer, "Monster Metaphors."

73. *Michael Jackson—Black Or White Official Music Video*, YouTube video, 5:06, posted by Jackson Putrus, June 1, 2012, https://www.youtube.com/watch?v=0RuAlyNclck

74. See for example this memorial article: Stanton Peele, "The Waste of Michael Jackson," *Psychology Today*, July 7, 2009, http://www.psychologytoday.com/blog/addiction-in-society/200907/the-waste-michael-jackson.

75. Jones, *Skintight*, 2008.

76. See for example Natalie Clarke, "Did Michael Jackson Want to Be White?" *Daily Mail Online*, June 30, 2009, http://www.dailymail.co.uk/news/article-1195843/Did-Michael-Jackson-want-white.html.

77. Jones, *Skintight*, 2008.

78. Ibid.
79. Davis, "Surgical Passing."
80. Zellweger's appearance generated coverage even in mainstream media outlets like the *New York Times*. Kuczynski, "Why the Strong Reaction to Renée Zellweger's Face?"
81. Perry, "Buying White Beauty."
82. Lauren Berlant, "Structures of Unfeeling: Mysterious Skin," *International Journal of Politics, Culture, and Society* 28, no. 3 (2015), 191–213, doi:10.1007/s10767-014-9190-y.
83. Butler, *Precarious Life*.
84. Though Lévinas would grant that once the face is placarded it ceases to be a face and becomes only a façade, which Butler builds upon to note that the universal face of x is employed as a marker. In being so deployed, it defaces the face itself.

BIBLIOGRAPHY

Abrams, Thomas. "Being-towards-Death and Taxes: Heidegger, Disability and the Ontological Difference." *Canadian Journal of Disability Studies* 2, no. 1 (2013): 28–50.

Agamben, Giorgio. *Homo Sacer: Sovereign Power and Bare Life*. Meridian. Stanford, CA: Stanford University Press, 1998.

Agich, George J., and Maria Siemionow. "Until They Have Faces: The Ethics of Facial Allograft Transplantation." *Journal of Medical Ethics* 31, no. 12 (2005): 707–9. doi:10.1136/jme.2005.011841.

Ahmed, Sara. *Strange Encounters: Embodied Others in Post-Coloniality*. London: Routledge, 2013.

Alexander, Shana. "They Decide Who Lives, Who Dies: Medical Miracle Puts Moral Burden on Small Committee." *Life*, November 9, 1962.

Allmark, Peter, and Su Mason. "Should Desperate Volunteers Be Included in Randomised Controlled Trials?" *Journal of Medical Ethics* 32, no. 9 (2006): 548–53. doi:10.1136/jme.2005.014282.Appadurai, Arjun. "Introduction: Commodities and the Politics of Value." In *The Social Life of Things: Commodities in Cultural Perspective*, edited by Arjun Appadurai, 3–63. Cambridge: Cambridge University Press, 1986.

Arendt, Hannah. *On Revolution*. London: Penguin Books, 1990.

Ashby, Imogen. "The Mutant Woman: The Use and Abuse of the Female Body in Performance Art." *Contemporary Theatre Review* 10, no. 3 (2000): 39–51. doi:10.1080/10486800008568595.

Ayers, Robert. "Serene and Happy and Distant: An Interview with Orlan." *Body & Society* 5, no. 2–3 (1999): 171–84. doi:10.1177/1357034X99005002010.

Balsamo, Anne. "On the Cutting Edge: Cosmetic Surgery and the Technological Production of the Gendered Body." *Camera Obscura* 10, no. 128 (1992): 206–37.

Barker, John H., Allen Furr, Michael Cunningham, Federico Grossi, Dalibor Vasilic, Barckley Storey, Osborne Wiggins, et al. "Investigation of Risk Acceptance in Facial Transplantation." *Plastic and Reconstructive Surgery* 118, no. 3 (2006): 663–70. doi:10.1097/01. prs.0000233202.98336.8c.

Baylis, Françoise. "A Face Is Not Just Like a Hand: Pace Barker." *American Journal of Bioethics* 4, no. 3 (2004): 30–32; discussion W23–31. doi:10.1080/15265160490496804.

Beauchamp, Tom. "The Principle of Beneficence in Applied Ethics." In *Stanford Encyclopedia of Philosophy*, edited by Edward N. Zalta. October 3, 2013. http://plato.stanford.edu/archives/win2013/entries/principle-beneficence/.

Bell-Metereau, Rebecca. "Stealth, Sexuality, and Culture Status in *The Manchurian Candidate* and *Seconds*." In *A Little Solitaire: John Frankenheimer and American Film*, edited by Murray Pomerance and R. Barton Palmer, 48–61. New Brunswick, NJ: Rutgers University Press, 2011.

Berlant, Lauren. "Structures of Unfeeling: Mysterious Skin." *International Journal of Politics, Culture, and Society*, 28, no. 3 (2015) 191–213. doi:10.1007/s10767-014-9190-y.

Bermudez, Luis. "Face Transplant: Is It Worth It?" *Plastic and Reconstructive Surgery* 117, no. 6 (2006): 1891–96. doi:10.1097/01.prs.0000219070.92470.0e.

Bivens, Rena. "The Gender Binary Will Not Be Deprogrammed: Ten Years of Coding Gender on Facebook." SSRN scholarly paper, Social Science Research Network, Rochester, NY, April 30, 2014. http://papers.ssrn.com/abstract=2431443.

Bluhm, Carla, and Nathan Clendenin. *Someone Else's Face in the Mirror: Identity and the New Science of Face Transplants*. Westport, CT: Praeger, 2009.

Bogart, Kathleen R., Linda Tickle-Degnen, and Nalini Ambady. "Communicating without the Face: Holistic Perception of Emotions of People with Facial Paralysis." *Basic and Applied Social Psychology* 36, no. 4 (2014): 309–20. doi:10.1080/01973533.2014.917973.

Bonilla-Silva, Eduardo. *Racism without Racists: Color-Blind Racism and the Persistence of Racial Inequality in the United States*. Lanham, MD: Rowman & Littlefield, 2006.

Bordo, Susan. *Unbearable Weight: Feminism, Western Culture, and the Body*. Berkeley: University of California Press, 2004.

Bosk, Charles. "Obtaining Voluntary Consent for Research in Desperately Ill Patients." *Medical Care* 40, no. 9 (2002): 64–68.

Bouchard, Gianna. "Incisive Acts: Orlan Anatomised." In *Orlan: A Hybrid Body of Artworks*, edited by Simon Donger, Simon Shepherd, and Orlan, 62–73. New York: Routledge, 2010.

Bourke, Joanna. "The Battle of the Limbs: Amputation, Artificial Limbs and the Great War in Australia." *Australian Historical Studies* 29, no. 110 (1998): 49–67. doi:10.1080/10314619808596060.

Bowen, Jon. "Gaining Face." *Salon*, May 19, 1999. http://www.salon.com/1999/05/19/face_transplants/.

Brown Clark, Stephanie. "Frankenflicks: Medicine and Monstrosity in Fiction Film." In *Cultural Sutures: Medicine and Media*, edited by Lester D. Friedman, 129–48. Durham, NC: Duke University Press, 2004.

Buber, Martin. *I and Thou*. Mansfield Centre, CT: Martino Fine Books, 2010.

Butler, Judith. *Precarious Life: The Powers of Mourning and Violence*. London: Verso, 2004.

Callahan, Daniel. "Bioethics as a Discipline." *Hastings Center Studies* 1 (1973): 66–73.

Caplan, Arthur. "Organ Transplantation." In *The Hastings Center Bioethics Briefing Book for Journalists, Policymakers, and Campaigns*, edited by Mary Crowley, 129–32. Garrison, NY: The Hastings Center, 2008.

Carmen, Marcela G. del, and Steven Joffe. "Informed Consent for Medical Treatment and Research: A Review." *Oncologist* 10, no. 8 (2005): 636–41. doi:10.1634/theoncologist.10-8-636.

Carosella, Edgardo D., and Thomas Pradeu. "Transplantation and Identity: A Dangerous Split?" *Lancet* 368, no. 9531 (2006): 183–84.

Chrysochou, Panayiota. "Orlan's Gruesome Theatre: Religion, Technology and the Politics of Female Self-Mutilation and Violence as Gestus." In *Female Beauty in Art: History, Feminism, Women Artists*, edited by Maria Ioannou and Maria Kyriakidou, 67–98. Newcastle upon Tyne: Cambridge Scholars, 2014.

Clark, David, and Catherine Myser. "Being Humaned: Medical Documentaries and the Hyperrealization of Conjoined Twins." In *Freakery: Cultural Spectacles of the Extraordinary Body*, edited by Rosemarie Garland Thomson, 338–55. New York: New York University Press, 1994.

Clark, David L., and Catherine Myser. "'Fixing' Katie and Eilish: Medical Documentaries and the Subjection of Conjoined Twins." *Literature and Medicine* 17, no. 1 (1998): 45–67. doi:10.1353/lm.1998.0004.

Coleman-Fountain, Edmund, and Janice McLaughlin. "The Interactions of Disability and Impairment." *Social Theory & Health* 11, no. 2 (2013): 133–50. doi:10.1057/sth.2012.21.

Colley, Linda. *Britons: Forging the Nation, 1707–1837*. Rev. ed. New Haven, CT: Yale University Press, 2009.

Cornea, Christine. "Frankenheimer and the Science Fiction/Horror Film." In *A Little Solitaire: John Frankenheimer and American Film*, edited by Murray Pomerance and R. Barton Palmer, 229–43. New Brunswick, NJ: Rutgers University Press, 2011.

Creed, Barbara. "Horror and the Monstrous-Feminine: An Imaginary Abjection." *Screen* 27, no. 1 (1986): 44–71. doi:10.1093/screen/27.1.44.

Crowe, Sam, Eric Cohen, and Alan Rubenstein. "Increasing the Supply of Human Organs: Three Policy Proposals." The President's Council on Bioethics. February, 2007. https://bioethicsarchive.georgetown.edu/pcbe/background/increasing_supply_of_human_organs.html.

Cypher, Joy M. "Agency and the Dominant Face: Facial Transplantation and the Discourse of Normalcy." *American Communication Journal* 12 (Winter 2010). http://ac-journal.org/journal/pubs/2010/ACJproofCypher.pdf.

Daston, Lorraine, and Katharine Park. *Wonders and the Order of Nature, 1150–1750*. New York: Zone Books, 1998.

Davis, Kathy. *Reshaping the Female Body: The Dilemma of Cosmetic Surgery*. New York: Routledge, 1995.

———. "Surgical Passing: Or Why Michael Jackson's Nose Makes 'Us' Uneasy." *Feminist Theory* 4, no. 1 (2003): 73–92.

Davis, Lennard J. "Constructing Normalcy." In *The Disability Studies Reader*, edited by Lennard J. Davis, 3rd ed., 9–28. New York: Routledge, 2010.

Deleuze, Gilles, and Felix Guattari. *A Thousand Plateaus: Capitalism and Schizophrenia*. Translated by Brian Massumi. Minneapolis: University of Minnesota Press, 1987.

Derrida, Jacques. *Archive Fever: A Freudian Impression*. Chicago: University of Chicago Press, 1996.

Dibbell, Julian. *My Tiny Life: Crime and Passion in a Virtual World*. New York: Holt Paperbacks, 1999.

Dittmann, Melissa. "Plastic Surgery: Beauty or Beast?" *Monitor on Psychology* 36, no. 8 (2005): 30.

Dorrian, Mark. "On the Monstrous and the Grotesque." *Word & Image* 16, no. 3 (2000): 310–17. doi:10.1080/02666286.2000.10435688.

Douglas, Mary. "No Free Gifts" Foreword to *The Gift: The Form and Reason for Exchange in Archaic Societies*, by Marcel Mauss, translated by W. D Halls. New York: WW Norton, 1990.

Drew-Bear, Annette. *Painted Faces on the Renaissance Stage: The Moral Significance of Face-Painting Conventions*. Lewisburg, PA: Bucknell University Press, 1994.

Durbach, Nadja. *Spectacle of Deformity: Freak Shows and Modern British Culture*. Berkeley: University of California Press, 2010.

Durham, Frank D. "News Frames as Social Narratives: TWA Flight 800." *Journal of Communication* 48, no. 4 (1998): 100–117. doi:10.1111/j.1460-2466.1998.tb02772.x.

Elliott, Carl, and Peter D. Kramer. *Better Than Well: American Medicine Meets the American Dream*. Rep. ed. New York: WW Norton, 2004.

Faber, Alyda. "Saint Orlan: Ritual as Violent Spectacle and Cultural Criticism." *TDR/The Drama Review* 46, no. 1 (2002): 85–92. doi:10.1162/105420402753555868.

Farrell, Colleen. "Telltale Purple Lesions." *Atrium: The Report of the Northwestern Medical Humanities and Bioethics Program* 9 (Spring 2011): 10–11.

Flannery, Mary M. "Research on the Terminally Ill: A Balancing Act Between Facilitating Access to Innovative Therapies and Protecting Vulnerable Subjects in Search of One Last Hope for Survival." Harvard University, 2003. http://nrs.harvard.edu/urn-3:HUL.InstRepos:8852203.

Foley, Robert N., and Raymond M. Hakim. "Why Is the Mortality of Dialysis Patients in the United States Much Higher than the Rest of the World?" *Journal of the American Society of Nephrology* 20, no. 7 (2009): 1432–35. doi:10.1681/ASN.2009030282.

Garg, Pushkal P., Susan L. Furth, Barbara A. Fivush, and Neil R. Powe. "Impact of Gender on Access to the Renal Transplant Waiting List for Pediatric and Adult Patients." *Journal of the American Society of Nephrology* 11, no. 5 (2000): 958–64.

Garland-Thomson, Rosemarie. "Feminist Disability Studies." *Signs* 30, no. 2 (2005): 1557–87.

———. "Integrating Disability, Transforming Feminist Theory." *NWSA Journal* 14, no. 3 (2002): 1–32.

———. *Staring: How We Look*. Oxford: Oxford University Press, 2009.

Geroulanos, Stefanos. "Postwar Facial Reconstruction: Georges Franju's *Eyes without a Face*." *French Politics, Culture & Society* 31, no. 2 (2013): 15–33.

Gever, Martha. "The Trouble with Moralism: Nip/Tuck." *Media, Culture & Society* 32, no. 1 (2010): 105–22.

Gilman, Sander L. *Creating Beauty to Cure the Soul: Race and Psychology in the Shaping of Aesthetic Surgery*. Durham, NC: Duke University Press, 1998.

———. *Making the Body Beautiful: A Cultural History of Aesthetic Surgery*. Princeton, NJ: Princeton University Press, 1999.

Gimlin, Debra. "Imagining the Other in Cosmetic Surgery." *Body & Society* 16, no. 4 (2010): 57–76. doi:10.1177/1357034X10383881.

Goffman, Erving. *The Presentation of Self in Everyday Life*. New York: Anchor, 1959.

Gonder, Patrick. "Like a Monstrous Jigsaw Puzzle: Genetics and Race in Horror Films of the 1950s." *Velvet Light Trap* 52, no. 1 (2003): 33–44. doi:10.1353/vlt.2003.0022.

Goodley, Dan. *Disability Studies: An Interdisciplinary Introduction*. London: Sage Publications, 2011.

Gordon, Chad R., Maria Siemionow, Kathy Coffman, Daniel Alam, Bijan Eghtesad, James E. Zins, Steven Bernard, John Fung, Landon Pryor, and Francis Papay. "The Cleveland Clinic FACES Score: A Preliminary Assessment Tool for Identifying the Optimal Face Transplant Candidate." *Journal of Craniofacial Surgery* 20, no. 6 (2009): 1969–74. doi:10.1097/SCS.0b013e3181bd2c86.

Gordon, Chad R., Edward W. Swanson, Srinivas M. Susarla, Devin Coon, Erin Rada, Mohammed Al Rakan, Gabriel F. Santiago, et al. "Overcoming Cross-Gender Differences and Challenges in Le Fort-Based, Craniomaxillofacial Transplantation with Enhanced Computer-Assisted Technology." *Annals of Plastic Surgery* 71, no. 4 (2013): 421–28. doi:10.1097/SAP.0b013e3182a0df45.

Govan, Emma. "Visceral Excess: Cosmimesis in the Work of Orlan." *Women & Performance* 15, no. 2 (2005): 147–59. doi:10.1080/07407700508571509.

Grealy, Lucy, and Ann Patchett. *Autobiography of a Face*. New York: Harper Perennial, 2003.

Habermas, Jürgen. *The Structural Transformation of the Public Sphere: An Inquiry into a Category of Bourgeois Society*. Rep. ed. Cambridge, MA: MIT Press, 1991.

Haiken, Elizabeth. *Venus Envy: A History of Cosmetic Surgery*. Baltimore: Johns Hopkins University Press, 1997.

Halpern, Orit. *Beautiful Data: A History of Vision and Reason Since 1945*. Experimental Futures. Durham, NC: Duke University Press, 2014.

Hamermesh, Daniel S. *Beauty Pays: Why Attractive People Are More Successful*. Princeton, NJ: Princeton University Press, 2011.

"Hand Transplant History." *Composite Tissue Allotransplantation*, April 12, 2015. http://www.handtransplant.com/TheProcedure/HandTransplantHistory/tabid/96/Default.aspx.

Haraway, Donna Jeanne. "A Cyborg Manifesto: Science, Technology, and Socialist-Feminism in the Late Twentieth Century." In *Simians, Cyborgs, and Women: The Reinvention of Nature*, 149–81. New York: Routledge, 1991.

Hartman, Rhonda Gay. "Face Value: Challenges of Transplant Technology." *American Journal of Law & Medicine* 31, no. 1 (2005): 7–46.

Hawkins, Joan. *Cutting Edge: Art-Horror and the Horrific Avant-Garde*. Minneapolis: University of Minnesota Press, 2000.

Hegel, G. W. F. *Phenomenology of Spirit*. Translated by A. V. Miller. Oxford: Oxford University Press, 1977.

Helfgot, Susan Whitman, and William Novak. *The Match: Complete Strangers, a Miracle Face Transplant, Two Lives Transformed*. New York: Simon & Schuster, 2013.

Herndon, April. "Disparate but Disabled: Fat Embodiment and Disability Studies." *NWSA Journal* 14, no. 3 (2002): 120–37.

Herschbach, Lisa. "Prosthetic Reconstructions: Making the Industry, Re-Making the Body, Modeling the Nation." *History Workshop Journal* Autumn, no. 44 (n.d.): 22–57, accessed July 13, 2016, http://hwj.oxfordjournals.org/content/1997/44/22.full.pdf.

Hettiaratchy, Shehan, and Peter E. M. Butler. "Face Transplantation—Fantasy or the Future?" *Lancet* 360, no. 9326 (2002): 5–6. doi:10.1016/S0140-6736(02)09361-3.

Hettiaratchy, Shehan, Mark A. Randolph, François Petit, W. P. Andrew Lee, and Peter E. M. Butler. "Composite Tissue Allotransplantation—A New Era in Plastic Surgery?" *British Journal of Plastic Surgery* 57, no. 5 (2004): 381–91. doi:10.1016/j.bjps.2004.02.012.

Heyes, Cressida J. *Self-Transformations: Foucault, Ethics, and Normalized Bodies*. Oxford: Oxford University Press, 2007.

Holliday, Ruth, and Jacqueline Sanchez Taylor. "Aesthetic Surgery as False Beauty." *Feminist Theory* 7, no. 2 (2006): 179–95. doi:10.1177/1464700106064418.

Hurley, Kelly. "Reading Like an Alien: Posthuman Identity in Ridley Scott's *Alien* and David Cronenberg's *Rabid*." In *Posthuman Bodies*, edited by Judith Halberstam and Ira Livingston, 203–24. Bloomington: Indiana University Press, 1995.

Introna, Lucas D. "The Face and the Interface: Thinking with Lévinas on Ethics and Justice in an Electronically Mediated World." Working paper, Department of Organisation, Work, and Technology, Lancaster: Lancaster University, 2003.

Jackson, John L. *Real Black: Adventures in Racial Sincerity*. Chicago: University of Chicago Press, 2005.

Jones, Jon W., Scott A. Gruber, John H. Barker, and Warren C. Breidenbach. "Successful Hand Transplantation—One-Year Follow-Up." *New England Journal of Medicine* 343, no. 7 (2000): 468–73. doi:10.1056/NEJM200008173430704.

Jones, Meredith. "Media-Bodies and Screen-Births: Cosmetic Surgery Reality Television." *Continuum* 22, no. 4 (2008): 515–24. doi:10.1080/10304310802189998.

———. *Skintight: An Anatomy of Cosmetic Surgery*. New York: Berg, 2008.

Jonsen, Albert R. *The Birth of Bioethics*. New York: Oxford University Press, 1998. http://site.ebrary.com/id/10375228.

Kael, Pauline. *5001 Nights at the Movies: A Guide from A to Z*. New York: Holt, Rinehart and Winston, 1982.

Kalat, David. "*Eyes without a Face*: The Unreal Reality," Criterion Collection, October 16, 2013, https://www.criterion.com/current/posts/343-eyes-without-a-face-the-unreal-reality.

Kalra, Gurvinder G., and Dinesh Bhugra. "Representation of Organ Transplantation in Cinema and Television." *International Journal of Organ Transplantation Medicine* 2, no. 2 (2011): 93–100.

Kaplan, Deborah. "The Definition of Disability: Perspective of the Disability Community." *Journal of Health Care Law and Policy* 3, no. 2 (2000): 352–64.

Kember, Sarah, and Joanna Zylinska. *Life after New Media: Mediation as a Vital Process*. Cambridge, MA: MIT Press, 2012.

Kendall, Lori. *Hanging Out in the Virtual Pub: Masculinities and Relationships Online.* Berkeley: University of California Press, 2002.

Khalifian, Saami, Philip S. Brazio, Raja Mohan, Cynthia Shaffer, Gerald Brandacher, Rolf N. Barth, and Eduardo D. Rodriguez. "Facial Transplantation: The First 9 Years." *Lancet* 384, no. 9960 (2014): 2153–63. doi:10.1016/S0140-6736(13)62632-X.

Khan, Firdos Alam. *Biotechnology Fundamentals.* Boca Raton, FL: CRC Press, 2012.

Kramer, Peter D. *Listening to Prozac: A Psychiatrist Explores Antidepressant Drugs and the Remaking of the Self.* New York: Viking Press, 1993.

Kruvand, Marjorie. "Face to Face: How the Cleveland Clinic Managed Media Relations for the First U.S. Face Transplant." *Public Relations Review* 36, no. 4 (2010): 367–75. doi:10.1016/j.pubrev.2010.09.001.

LaFrance, Marc. "'She Exists within Me': Subjectivity, Embodiment, and the World's First Face Transplant." In *Abjectly Boundless: Boundaries, Bodies and Health Work*, edited by Trudy Rudge and Dave Holmes, 147–59. Farnham: Ashgate, 2010.

Laskas, Jeanne Marie. "The Miraculous Face Transplant of Richard Norris." *GQ*, August 28, 2014. http://www.gq.com/news-politics/newsmakers/201408/richard-norris.

Laurel, Brenda. *Computers as Theatre.* Upper Saddle River, NJ: Pearson Education, 2013.

Leach, Gerald. *The Biocrats: Implications of Medical Progress.* London: Cape, 1970.

Lederer, Susan E. *Flesh and Blood: Organ Transplantation and Blood Transfusion in Twentieth-Century America.* Oxford: Oxford University Press, 2008.

Lengelé, Benoît, Sylvie Testelin, Sophie Cremades, and Bernard Devauchelle. "Facing Up Is an Act of Dignity: Lessons in Elegance Addressed to the Polemicists of the First Human Face Transplant." *Plastic and Reconstructive Surgery* 120, no. 3 (2007): 803–6. doi:10.1097/01.prs.0000271097.22789.79.

Leverette, Marc. "The Funk of Forty Thousand Years; Or, How the (Un)Dead Get Their Groove On." In *Zombie Culture: Autopsies of the Living Dead*, edited by Shawn McIntosh and Marc Leverette, 185–212. Lanham, MD: Scarecrow Press, 2008.

Lévinas, Emmanuel. *Ethics and Infinity: Conversations with Philippe Nemo.* Translated by Richard A. Cohen. Pittsburgh: Duquesne University Press, 1985.

———. *Totality and Infinity: An Essay on Exteriority.* The Hague: Martinus Nijhoff, 1979.

———. "The Trace of the Other." In *Deconstruction in Context: Literature and Philosophy*, edited by Mark C. Taylor, 345–59. Chicago: University of Chicago Press, 1986.

Lévinas, Emmanuel, and Philippe Nemo. *Ethics and Infinity.* Pittsburgh: Duquesne University Press, 1985.

Lewis, Tania. "Revealing the Makeover Show." *Continuum* 22, no. 4 (2008): 441–46. doi:10.1080/10304310802190053.

———. *Smart Living: Lifestyle Media and Popular Expertise.* New York: Peter Lang, 2008.

Liesch, John, and Cindy Patton. "Clinic or Spa? Facial Surgery in the Context of AIDS-Related Facial Wasting." In *The Rebirth of the Clinic*, edited by Cindy Patton, 1–16. Minneapolis: University of Minnesota Press, 2010.

Lingel, Jessa, and Tarleton Gillespie. "One Name to Rule Them All: Facebook's Identity Problem." *The Atlantic*, October 2, 2014. http://www.theatlantic.com/technology/archive/2014/10/one-name-to-rule-them-all-facebook-still-insists-on-a-single-identity/381039/.

Linker, Beth. "On the Borderland of Medical and Disability History: A Survey of the Fields." *Bulletin of the History of Medicine* 87, no. 4 (2013): 499–535. doi:10.1353/bhm.2013.0074.

———. *War's Waste: Rehabilitation in World War I America*. Chicago: University of Chicago Press, 2011.

Linton, Simi. *Claiming Disability: Knowledge and Identity*. New York: New York University Press, 1998.

Lock, Margaret M. *Twice Dead: Organ Transplants and the Reinvention of Death*. California Series in Public Anthropology 1. Berkeley: University of California Press, 2002.

Lodge, David. *Changing Places: A Tale of Two Campuses*. 2nd ed. New York: Penguin Books, 1979.

Longmore, Paul K. "Screening Stereotypes: Images of Disabled People." In *Screening Disability: Essays on Cinema and Disability*, edited by Christopher R. Smit and Anthony Enns, 1–17. Lanham, MD: University Press of America, 2001.

Lowenstein, Adam. *Shocking Representation: Historical Trauma, National Cinema, and the Modern Horror Film*. Film and Culture. New York: Columbia University Press, 2005.

Magnus, David, Paul Root Wolpe, Kelly Carroll, and Glenn McGee. "A New Era for *AJOB*." *American Journal of Bioethics* 4, no. 3 (2004): x–xi.

Marchetti, Clelia, and Sergio Della Sala. "Disentangling the Alien and Anarchic Hand." *Cognitive Neuropsychiatry* 3, no. 3 (1998): 191–207. doi:10.1080/135468098396143.

Marcus, Sharon, ed. "Special Issue on Celebrities and Publics in the Internet Era." *Public Culture* 27, no. 1 (2015).

Marvin, Carolyn. *Blood Sacrifice and the Nation: Totem Rituals and the American Flag*. Cambridge Cultural Social Studies. Cambridge: Cambridge University Press, 1999.

———. "Communication as Embodiment." In *Communication As . . . : Perspectives on Theory*, edited by Gregory J. Shepherd, Jeffrey St. John, and Theodore G. Striphas, 67–74. Thousand Oaks, CA: Sage Publications, 2006.

Mauss, Marcel. *The Gift: The Form and Reason for Exchange in Archaic Societies*. Translated by W. D Halls. New York: WW Norton, 1990.

McGrath, Patrick. "Appearances to the Contrary: *Eyes without a Face*." Criterion Collection, July 20, 2014. http://www.criterion.com/current/posts/677-appearances-to-the-contrary-eyes-without-a-face.

McQuire, Scott. "From Glass Architecture to Big Brother: Scenes from a Cultural History of Transparency." *Cultural History Review* 9, no. 1 (2003): 103–23.

Mercer, Kobena. "Monster Metaphors: Notes on Michael Jackson's 'Thriller.'" *Screen* 27, no. 1 (1986): 26–43. doi:10.1093/screen/27.1.26.

Mihm, Stephen. "'A Limb Which Shall Be Presentable in Polite Society': Prosthetic Technologies in the Nineteenth Century." In *Artificial Parts, Practical Lives: Modern Histories of Prosthetics*, 282–99. New York: New York University Press, 2002.

Miranda, Krista. "DNA, AND: A Meditation on Pandrogeny." *Women & Performance: A Journal of Feminist Theory* 20, no. 3 (2010): 347–53. doi:10.1080/07407 70X.2010.529266.

Mirowski, Philip, and Dieter Plehwe, eds. *The Road from Mont Pèlerin: The Making of the Neoliberal Thought Collective*. Cambridge, MA: Harvard University Press, 2009.

Mock, Roberta. "Stand-up Comedy and the Legacy of the Mature Vagina." *Women & Performance: A Journal of Feminist Theory* 22, no. 1 (2012): 9–28. doi:10.1080/07407 70X.2012.685394.

Mules, Warwick. "This Face: A Critique of Faciality as Mediated Self-Presence." The Face and Technology *Transformations* no. 18 (2010).

Nakamura, Lisa. *Cybertypes: Race, Ethnicity, and Identity on the Internet*. New York: Routledge, 2002.

Nathoo, Ayesha. *Hearts Exposed: Transplants and the Media in 1960s Britain*. Science, Technology, and Medicine in Modern History. Basingstoke: Palgrave Macmillan, 2009.

Naugler, Diane. "Crossing the Cosmetic/Reconstructive Divide: The Instructive Situation of Breast Reduction Surgery." In *Cosmetic Surgery: A Feminist Primer*, edited by Cressida J. Heyes and Meredith Jones, 224–37. Farnham: Ashgate, 2009.

Nisbet, Matthew C. "The Competition for Worldviews: Values, Information, and Public Support for Stem Cell Research." *International Journal of Public Opinion Research* 17, no. 1 (2005): 90–112. doi:10.1093/ijpor/edh058.

O'Bryan, C. Jill. *Carnal Art: Orlan's Refacing*. Minneapolis: University of Minnesota Press, 2005.

Olney, Ian. "The Problem Body Politic, or 'These Hands Have a Mind All Their Own!': Figuring Disability in the Horror Film Adaptations of Renard's *Les Mains d'Orlac*." *Literature/Film Quarterly* 34, no. 4 (2006): 294–302.

O'Neill, Robert D. "'Frankenstein to Futurism': Representations of Organ Donation and Transplantation in Popular Culture." *Transplantation Reviews* 20, no. 4 (2006): 222–30. doi:10.1016/j.trre.2006.09.002.

Parameswaran, Radhika. "Global Queens, National Celebrities: Tales of Feminine Triumph in Post-liberalization India." *Critical Studies in Media Communication* 21, no. 4 (2004): 346–70. doi:10.1080/0739318042000245363.

Peacock, Louise. "Joan Rivers—Reading the Meaning." *Comedy Studies* 2, no. 2 (2011): 125–37. doi:10.1386/cost.2.2.125_1.

Pearl, Sharrona. *About Faces: Physiognomy in Nineteenth-Century Britain*. Cambridge, MA: Harvard University Press, 2010.

———. "Believing in Not Seeing: Teaching Atrocity without Images." *Afterimage* 40, no. 6 (2013): 16–20.

———. "Victorian Blockbuster Bodies and the Freakish Pleasure of Looking." *Nineteenth-Century Contexts* 38, no. 1 (forthcoming).

Pearl, Sharrona, and Alexandra Sastre. "The Image Is (Not) the Event: Negotiating the Pedagogy of Controversial Images." *Visual Communication Quarterly* 21, no. 4 (2014): 198–209. doi:10.1080/15551393.2014.987283.

Peele, Stanton. "The Waste of Michael Jackson." *Psychology Today*, July 7, 2009. http://www.psychologytoday.com/blog/addiction-in-society/200907/the-waste-michael-jackson.

Peiss, Kathy Lee. *Hope in a Jar: The Making of America's Beauty Culture*. Philadelphia: University of Pennsylvania Press, 2011.

Perry, Imani. "Buying White Beauty." *Cardozo Journal of Law & Gender* 12 (2006): 579–607.

Peters, John Durham. *Speaking into the Air: A History of the Idea of Communication*. Chicago: University of Chicago Press, 2001.
Petit, François, Antonis Paraskevas, Alicia B. Minns, W. P. Andrew Lee, and Laurent A. Lantieri. "Face Transplantation: Where Do We Stand?" *Plastic and Reconstructive Surgery* 113, no. 5 (2004): 1429–33.
Pick, Daniel. *Svengali's Web: The Alien Enchanter in Modern Culture*. New Haven, CT: Yale University Press, 2000.
Pinchevski, Amit. "Ethics on the Line." *Southern Communication Journal* 68, no. 2 (2003): 152–66. doi:10.1080/10417940309373257.
———. "Lévinas as a Media Theorist: Toward an Ethics of Mediation." *Philosophy and Rhetoric* 47, no. 1 (2014): 48–72.
Pinchevski, Amit, and John Durham Peters. "Autism and New Media: Disability between Technology and Society." *New Media & Society*, 2015, 1–17. doi:10.1177/1461444815594441.
Pitts-Taylor, Victoria. *Surgery Junkies: Wellness and Pathology in Cosmetic Culture*. New Brunswick, NJ: Rutgers University Press, 2007.
P-Orridge, Genesis. *Painful but Fabulous: The Life and Art of Genesis P-Orridge*. New York: Soft Skull Press, 2002.
Powell, Tia. "Face Transplant: Real and Imagined Ethical Challenges." *Journal of Law, Medicine & Ethics* 34, no. 1 (2006): 111–15. doi:10.1111/j.1748-720X.2006.00014.x.
Pratley, Gerald. *The Films of Frankenheimer: Forty Years in Film*. Bethlehem, PA: Lehigh University Press, 1998.
Prendergast, Mary B., and Robert S. Gaston. "Optimizing Medication Adherence: An Ongoing Opportunity to Improve Outcomes after Kidney Transplantation." *Clinical Journal of the American Society of Nephrology: CJASN* 5, no. 7 (2010): 1305–11. doi:10.2215/CJN.07241009.
Quandt, James. "*The Face of Another*: Double Vision," Criterion Collection, July 9, 2007, https://www.criterion.com/current/posts/592-the-face-of-another-double-vision
.Rabinow, Paul, and Nikolas Rose. "Biopower Today." *BioSocieties: An Interdisciplinary Journal for the Social Study of the Life Sciences* 1, no. 2 (2006): 195–217. doi:10.1017/S1745855206040014.
Reeve, Donna. "Cyborgs, Cripples and iCrip: Reflections on the Contribution of Haraway to Disability Studies." In *Disability and Social Theory*, edited by Dan Goodley, Bill Hughes, and Lennard Davis, 91–111. New York: Palgrave Macmillan, 2012.
Rembis, Mark A. "Disability Studies." In *International Encyclopedia of Rehabilitation*, edited by J. H. Stone and Maurice Blouin, 2010. http://cirrie.buffalo.edu/encyclopedia/en/article/281/.
Rheingold, Howard. *The Virtual Community: Homesteading on the Electronic Frontier*. Cambridge, MA: MIT Press, 1993.
Rose, David. *Enchanted Objects: Design, Human Desire, and the Internet of Things*. New York: Simon and Schuster, 2014.
Roth, Yoel. "'No Overly Suggestive Photos of Any Kind': Content Management and the Policing of Self in Gay Digital Communities." *Communication, Culture & Critique*, 8, no. 3 (2015), 414–32. doi:10.1111/cccr.12096.

Sarwer, David B., and Canice E. Crerand. "Body Dysmorphic Disorder and Appearance Enhancing Medical Treatments." *Body Image* 5, no. 1 (2008): 50–58. doi:10.1016/j.bodyim.2007.08.003.
Saunders, Robert. "Zombies in the Colonies: Imperialism and Contestation of Ethno-Political Space in Max Brooks' *The Zombie Survival Guide*." In *Monstrous Geographies: Places and Spaces of the Monstrous*, edited by Sarah Montin and Evelyn Tsitas. Oxford: Inter-Disciplinary Press, 2013, 19–46.
Schweik, Susan M. *The Ugly Laws: Disability in Public*. New York: New York University Press, 2010.
Sekula, Allan. "The Body and the Archive." *October* 39 (1986): 3–64.
Sender, Katherine. *The Makeover: Reality Television and Reflexive Audiences*. Critical Cultural Communication. New York: New York University Press, 2012.
Sennett, Richard. *The Fall of Public Man*. New York: WW Norton, 1992.
Serlin, David. *Replaceable You: Engineering the Body in Postwar America*. Chicago: University Of Chicago Press, 2004.
Sessions, Lauren F. "'You Looked Better on MySpace': Deception and Authenticity on the Web 2.0." *First Monday* 14, no. 7 (2009). doi:10.5210/fm.v14i7.2539.
Sharp, Lesley A. "Organ Transplantation as a Transformative Experience: Anthropological Insights into the Restructuring of the Self." *Medical Anthropology Quarterly* 9, no. 3 (1995): 357–89. doi:10.1525/maq.1995.9.3.02a00050.
Shell, Marc. "Moses' Tongue." *Common Knowledge* 12, no. 1 (2006): 150–76.
Shildrick, Margrit. "Corporeal Cuts: Surgery and the Psycho-Social." *Body & Society* 14, no. 1 (2008): 31–46. doi:10.1177/1357034X07087529.
———. "'Why Should Our Bodies End at the Skin?': Embodiment, Boundaries, and Somatechnics." *Hypatia* 30, no. 1 (2015): 13–29. doi:10.1111/hypa.12114.
Siebers, Tobin. *Disability Aesthetics*. Corporealities: Discourses of Disability. Ann Arbor: University of Michigan Press, 2010.
Siemionow, Maria, Sakir Unal, Galip Agaoglu, and Alper Sari. "A Cadaver Study in Preparation for Facial Allograft Transplantation in Humans: Part I. What Are Alternative Sources for Total Facial Defect Coverage?" *Plastic and Reconstructive Surgery* 117, no. 3 (2006): 864–72; discussion 873–75. doi:10.1097/01.prs.0000204875.10333.56.
Silk, Michael L., Jessica Fancombe, and Faye Bachelor. "The Biggest Loser: The Discursive Constitution of Fatness." *Interactions: Studies in Communication & Culture* 1, no. 3 (2011): 369–89. doi:10.1386/iscc.1.3.369_1.
Silverstone, Roger. *Media and Morality: On the Rise of the Mediapolis*. Cambridge: Polity, 2006.
Smart, Barry. "Facing the Body—Goffman, Lévinas and the Subject of Ethics." *Body & Society* 2, no. 2 (1996): 67–78. doi:10.1177/1357034X96002002004.
Smeets, Ralf, Carsten Rendenbach, Moritz Birkelbach, Ahmed Al-Dam, Alexander Be, Henning Hanken, and Max Heiland. "Face Transplantation: On the Verge of Becoming Clinical Routine?" *BioMed Research International* (June 9, 2014): e907272. doi:10.1155/2014/907272.
Sobchack, Vivian Carol. *Screening Space: The American Science Fiction Film*. 2nd. ed. Brunswick, NJ: Rutgers University Press, 1997.

Sothern, Matthew, and Jen Dickinson. "Repaying the Gift of Life: Self-Help, Organ Transfer and the Debt of Care." *Social & Cultural Geography* 12, no. 8 (2011): 889–903. doi:10.10 80/14649365.2011.624192.

"Special Issue on Face Transplantation." *American Journal of Bioethics* 4, no. 3 (2004).

Sperati, Giorgio. "Amputation of the Nose throughout History." *Acta Otorhinolaryngologica Italica* 29, no. 1 (2009): 44–50.

Stone, Allucquere Rosanne. "Will the Real Body Please Stand Up?: Boundary Stories about Virtual Cultures." In *Cyberspace: First Steps*, edited by Michael L. Benedikt, 81–118. Cambridge, MA: MIT Press, 1991.

Sudnow, David. *Passing On: The Social Organization of Dying*. Englewood Cliffs, NJ: Prentice-Hall, 1967.

Swift, Teresa. "Desperation May Affect Autonomy but Not Informed Consent." *American Journal of Bioethics Neuroscience* 2, no. 1 (2011).

Talley, Heather Laine. *Saving Face: Disfigurement and the Politics of Appearance*. New York: New York University Press, 2014.

Tandoc, Edson C. "Journalism Is Twerking? How Web Analytics Is Changing the Process of Gatekeeping." *New Media & Society* 16, no. 4 (2014): 559–75. doi:10.1177/ 1461444814530541.

Taylor-Alexander, Samuel. *On Face Transplantation: Life and Ethics in Experimental Biomedicine*. New York: Palgrave Macmillan, 2014.

The Royal College of Surgeons. "Facial Transplantation Working Party Report," November 2003. http://www.rcseng.ac.uk/publications/docs/facial_transplantation .html/@@download/pdffile/facial%20transplantation%202003.pdf.

"The Telltale Heart." *Ebony*, March 1968.

Thomas, Abraham, Vijay Obed, Anil Murarka, and Gopal Malhotra. "Total Face and Scalp Replantation." *Plastic and Reconstructive Surgery* 102, no. 6 (1998): 2085–87.

Tobin, G. R., W. C. Breidenbach, M. M. Klapheke, F. R. Bentley, D. J. Pidwell, and P. D. Simmons. "Ethical Considerations in the Early Composite Tissue Allograft Experience: A Review of the Louisville Ethics Program." *Transplantation Proceedings* 37, no. 2 (2005): 1392–95. doi:10.1016/j.transproceed.2004.12.179.

Tuchman, Gaye. "Telling Stories." *Journal of Communication* 26, no. 4 (1976): 93–97. doi:10.1111/j.1460-2466.1976.tb01942.x.

Turkle, Sherry. *Life on the Screen*. New York: Simon and Schuster, 2011.

Turner, Fred. *The Democratic Surround: Multimedia & American Liberalism from World War II to the Psychedelic Sixties*. Chicago: The University of Chicago Press, 2013.

Turner, Graeme. *Understanding Celebrity*. Thousand Oaks, CA: Sage Publications, 2013.

van Dijck, José. *The Transparent Body: A Cultural Analysis of Medical Imaging*. Seattle: University of Washington Press, 2005.

Vercler, Christian J. "Ethical Issues in Face Transplantation." *Virtual Mentor* 12, no. 5 (2010): 378. doi:10.1001/virtualmentor.2010.12.5.jdsc1-1005.

Volk, Michael L., Scott W. Biggins, Mary Ann Huang, Curtis K. Argo, Robert J. Fontana, and Renee R. Anspach. "Decision Making in Liver Transplant Selection Committees: A Multicenter Study." *Annals of Internal Medicine* 155, no. 8 (2011): 503–8. doi:10.7326/0003-4819-155-8-201110180-00006.

Warne, Vanessa. "'To Invest a Cripple with Peculiar Interest': Artificial Legs and Upper-Class Amputees at Mid-Century." *Victorian Review* 35, no. 2 (2009): 83–100.

Watts, Duncan J. *Six Degrees: The Science of a Connected Age*. New York: Norton, 2003.

Wegenstein, Bernadette. *The Cosmetic Gaze: Body Modification and the Construction of Beauty*. Cambridge, MA: MIT Press, 2012.

Wegenstein, Bernadette, and Nora Ruck. "Physiognomy, Reality Television and the Cosmetic Gaze." *Body & Society* 17, no. 4 (2011): 27–54. doi:10.1177/1357034X11410455.

Weiner, Annette B. *Inalienable Possessions: The Paradox of Keeping while Giving*. Berkeley: University of California Press, 1990.

Wilde, Oscar. *The Picture of Dorian Gray*. Dover Thrift Editions. New York: Dover Publications, 1993.

Wilson, David. "Review of *Seconds*." *Sight and Sound* 36, no. 1 (1966): 46.

Winter, Alison. *Mesmerized: Powers of Mind in Victorian Britain*. Chicago: University of Chicago Press, 1998.

Young, Allan. *The Harmony of Illusions: Inventing Post-Traumatic Stress Disorder*. Princeton, NJ: Princeton University Press, 1997.

Young, Lynne, and Maureen Little. "Women and Heart Transplantation: An Issue of Gender Equity?" *Health Care for Women International* 25, no. 5 (2004): 436–53. doi:10.1080/07399330490272778.

Zimmermann, Anja. "'Sorry for Having to Make You Suffer': Body, Spectator, and the Gaze in the Performances of Yves Klein, Gina Pane, and Orlan." *Discourse* 24, no. 3 (2002): 27–46. doi:10.1353/dis.2003.0035.

Zylinska, Joanna, and Gary Hall. "Probings: An Interview with Stelarc." In *The Cyborg Experiments: The Extensions of the Body in the Media Age*, edited by Joanna Zylinska, 114–30. New York: Bloomsbury Academic, 2002.

INDEX

Abe, Kobo, 59, 203n25
Abrams, Thomas, 19
Aetna, 147
Afghanistan, 151
Agamben, Giorgio, 66, 150
agency, and other, 180
Agich, George J., 108
Ahmed, Sara, 93
AIDS, 15
Alexander, Shana, 36
Algeria, 53
Algerian War, 204n42
American Journal of Bioethics (journal), 19, 111–12
American Journal of Law and Medicine (journal), 112
America's Top Model (reality television show), 30
Annals of Plastic Surgery (journal), 148
Appadurai, Arjun, 34, 67, 203n34
Arendt, Hannah, 66, 80, 87
Arias, Lorrie, 129, 196n61
Armed Forces Institute of Regenerative Medicine, 151
Arnold, Newton, 44
Australia, 11, 91, 97, 105, 206n26
authenticity, 30–31

bariatric surgery, 21
Barnard, Christian, 35
Bartlett, Stephen T., 151–52
Baylis, Françoise, 94, 109
Benson, Ryan, 129
Berlant, Lauren, 179; deadpan, analysis of, 180
Bermudez, Luis, 114–15
Bian Que, 32
Biggest Loser, The (reality television show), 128, 212m7
bioethics, 25, 31–32, 47, 102; informed consent, 199n107; as term, 37
Black or White (video), 175
Blaiberg, Philip, 42
bodily boundaries, queering of, 171, 174
bodily essentialism, 179
bodily integrity, 41; and fear, 47; identity, connected to, 43
bodily manipulation, 8, 14
bodily narrative, 28; difference, as medical problem, 29
body, 8, 178; changing nature of, 20; as contested space, 44; face, relationship between, 18–19; and history, 160; as index, 15; manipulation of, 115; materiality of, 156; mind, dualism between, 17–18, 46; as text, 15

body anxiety, 13
body dysmorphia, 23
body horror films, 44, 52
body rebellion films, 43
Bordo, Susan, 16, 161, 169, 171
Boutella, Sofia, 170
Bradbury, Eileen, 101
Brigham and Women's Hospital, 133, 135, 138, 149, 151–52, 214n37
Britain, 3. *See also* England; United Kingdom
Brown Clark, Stephanie, 45–46, 48
Buber, Martin, 166–67
Burch, Nöel, 60
Bussey, Georgia, 192n8
Butler, Judith, 2, 159, 180, 221n84
Butler, Peter E. M., 106–7, 206n26

Cage, Nicolas, 1
Callahan, Dan, 37
Caplan, Arthur, 93, 205n12
Carre, Jeff, 202n23
Carrell, Alexis, 33
Cartesian dualism, 173
catfishing, 217n20
cellular memory, identity transfer, 32, 38–42
Centre Hospitalier Universitaire Nord, 204n1
Chang, JuJu, 139
Changing Faces (documentary), 192n8
Changing Places (Lodge), 1
Cheng and Eng, 197n81
China, 105, 146–47
Chomsky, Noam, 96
Clark, David L., 130–31
cleft palate surgery, 192n7
Cleveland Clinic, 89, 95, 97, 103, 105, 107, 118, 136, 204n1, 206n26, 214n37
clickbait, 93, 117; as defined, 205n11; and flak, 96
Coleman-Fountain, Edmund, 19
Colley, Linda, 22
congenital abnormalities, as portents, 31
Cornea, Christine, 51
Cosmas, Saint, 32

cosmetic manipulation, 16
cosmetic surgery, 6–7, 13–14, 18, 38, 46–47, 94, 97, 102, 129, 154, 171, 194n28; after portion, 127; bioethical questions, 25; different look, 22; Down's Syndrome, 192n8; mental-health benefits of, 17; reality television, 124. *See also* makeover; plastic surgery
Crawford, Cindy, 198n84
Creed, Barbara, 65
Crerand, Canice, 23
Culp, Connie, 97–98, 107, 133–39, 148
Curry, Ann, 141–43
cyberspace, 164
cyborg, 15–16, 176; as goddess, 5, 169–70; and monsters, 5, 121–22
"Cyborg Manifesto, The" (Haraway), 15–16
Cypher, Joy, 95

Damian, Saint, 32
Dark Passage (film), 55
Davis, Kathy, 17, 23, 169, 174–76
Davis, Lennard, 44
death, 76; definition of, 37
Deleuze, Gilles, 2, 158–59, 168
Dent, Harvey "Two-Face" (character), 217n12
Department of Defense, 143, 151–52
Devauchelle, Bernard, 89
difference: and disfigurement, 5
digital avatars, 9
digital experience, play and performance, 161
digital self, 161
digital space, 217n21
Dinoire, Isabelle, 7, 10, 49, 52, 87, 107, 126, 128, 132–35, 147–48, 153–54, 172, 178, 200n121, 204n1, 206n26, 213n32; attempted suicide of, 88, 92, 97, 100; backlash against, 95–96; and clickbait, 93; cosmetic frame, employment of toward, 119; facial transplant surgery of, 89–90, 93, 95–102, 104–5; framing practices, 97; gratitude of, 117; immunosuppressants,

lifelong need for, 89–90; life story, controversy over, 118–19; masking of, 100; media coverage of, 95–102, 104–5, 117–18; and morality, 100; as outlier, 146; psychology of, 100–101, 118–19; quality-of-life issues, 90–92; sense of self, 96; surgery, condemnation of, 93, 97, 99, 115; therapeutic value on, 116; as undeserving, 101; as victim, 95
disability, 5, 176; definition of, 195n42
disability studies, 13, 19–20, 22–23, 47, 195n48
disability theory, 28
disfigurement, 3, 15, 24, 153, 176–77; acid attacks, 148; and difference, 5; as disability, 19–20, 117; as social death, 18
distance, and relationality, 9
Doctors, The (television reality show), 144
donations, 34, 41–42, 62, 133
donors, 34, 108, 134–39, 147–48
Dorrian, Mark, 43
doubling, 51, 55, 57–58, 82–84
Douglas, Mary, 34
Drew-Bear, Annette, 16, 203n39
Dr. Strangelove (film), 200n131
Dubernard, Jean-Michel, 89, 99–100, 102
Durbach, Nadja, 30–31
Durham, Frank, 96–97
dystopia, 49

Ebony (magazine), 41
Ecuador, 102
Ely, David, 58
England, 56, 91. *See also* Britain; United Kingdom
ESPN (television channel), 144–45
ethics, 34, 173; of face, 158; and self, 158
Evans, Danielle, 30
Extreme Makeover (reality television series), 176, 212n7
Eyes without a Face (film), 7, 49–50, 52, 55, 63–64, 68–69, 201n3, 202n14; deaths in, as punishment, 76, 80; description of, 56–57; doubling in, motif of, 57–58, 82–83; and facelessness, 85; family, playing with

idea of, 74–75; gifting framework in, 67; as horror film, 65; masking in, 75, 80, 82; medicine, dangers of in, 67, 70; morality in, 81; operation, institutional motif of in, 54; operation in, 65–66; replacement, as story of, 54; transformation in, 61; transportation motif in, 53; victimhood, concern over, 75; as World War II allegory, 74

face, 127; as absent, 180; as abstraction, 177; act of becoming, 4; of another, 63; and avatars, 164; as celebrity, 123; change, possibility of, 171; as changing, 2, 4, 55, 164, 173, 178; and character, 13–14, 24–25, 55; character, indexical relationship between, 2–3; and cheating, 27–28, 86; codedness of, 159–60; corporeality of, 161; and difference, 179; different kinds of, 4; and doubling, 84; emotional resonances of, 92; encounter with, 167; of faceless, 168; facial erasure, and history, 25–26; as flexible, 167; history of, and racism, 168; humanity, understanding of, 85, 180; human value of, 112; and identity, 6, 82, 94, 119; importance of, 156; as index, 3, 12, 24, 177; index, vexing of, 61, 87, 161; indexical nature of, 4, 8, 50, 104, 160–62, 180; loss of, and loss of character, 12; as lying, 27; making of, 161; manipulation of, 52, 160, 178; manipulation of, and horror films, 84–85; meaning of, 2, 5, 8; as multifaced, 160; mutilation of, 27; new ethics of, 8; new identity, 2, 13, 64, 124–25, 133; nuance of, 180; as other, 159–60, 167, 177; in process, 161; psychology of, 93–94; as public, 4, 164; public life, as passport to, 67–68; as relational, 4; self, as changing, 3; sense of self, 71, 160; signification, iconic form of, 159–60; special status of, 13–14, 92, 94, 102, 206n20, 209n65; specificity of, 171; symbolic significance of, 109; transformation of, 52; transplantation of, 52; as unfixed, 4

face-bodies, 161
Facebook, 205n11; bootstrapping, process as, 162
facehood, as nonindexical, 156
Face in the Crowd, A (film), 55
faceless, 168, 179–80; face-to-face encounter, 3; as left for dead, 3–4; as monsters, 52, 178; and online, 163
Faceless (film), 55
facelessness, 8, 21, 29, 72, 80–81, 87, 91, 95, 130, 139–40, 148, 155, 159–60, 209n65; condition of, 3; conjoined twins, comparison between, 131; as contextual, 20; and death, 76; and doubling, 84; rebirth, earning of, 128; as stigma, 20; trauma of, 2, 109; of veterans, 150
Face of Another, The (film), 7, 49–50, 52, 69, 201n3, 202n14; deaths in, as punishment, 76; description of, 59–61; donor-patient relationship in, 62; doubling in, motif in, 82–83; face, attempt to regain, 72; and facelessness, 85; masking in, 53, 56–62, 64, 70–74, 78–80, 82–84; medicine, dangers of in, 67, 70–72; operation, institutional motif of in, 54; punishment motif in, 78–80; replacement, as story of, 54; secondary plot of, 72–73; space of exchange in, 63; transformation in, 61; transportation motif in, 53
Face/Off (film), 1, 55
"face race," 7, 92, 102, 206n26
"Face Transplantation—Fantasy or the Future?" (Hettiaratchy and Butler), 106
"Face Transplant: Real and Imagined Ethical Challenges" (Powell), 111
face-transplant recipients, 176, 178; after portion, 127; before portion, 127; big reveal, 129, 133, 137, 139, 141, 156; as celebrities, 30, 123, 130, 132, 153; corporeality of, 130–31; as deserving, 101, 128, 153–54; doctors, as makeover specialists, 154; failure, possibility of, 129–30, 133; gendered expectations of, 146–49; gratitude of, 132, 136–37, 139, 141, 144; makeover of, 123–27, 131; as new self, 154; patienthood of, 130; public personas of, 131; scars, as indexical of experience, 124, 126; as special makeover, 152; structure of, 132–33, 144; television portrayals of, 123–24; time, changing of, 154; transformation of, 126–27, 130, 132, 134, 138–40, 143, 145, 150, 153–54; transgressive state of being of, 131; trauma of, 124–25, 127; trauma of, as ennobling, 128, 139, 145, 154; as work in progress, 127–28. *See also* makeover; *and individual recipients*
face-transplant surgery, 1, 3, 10, 12, 19–21, 25–26, 32, 37–38, 43, 61–62, 168, 177, 205n5, 206n20, 208n58, 209n65, 210n94; ambiguity of, 47, 91, 155; and anger, 49; anonymity, specter of, 55; appearance, of another, 24, 47; as becoming, 172–73; bioethics of, 91–95, 97, 99–100, 102, 104–6, 111–12, 116, 119, 122; as cheating, 2, 99, 121; controversy over, 103; cosmetic surgery, 2, 5, 17–18, 95, 98–100, 102, 104–5, 117; as dangerous, 99; as discomfiting, 173; doctors, on trial, 102, 104–5, 121; and donors, 136–37, 141, 151, 204n3, 204–5n4, 215n64; edutainment, as form of, 132; as elective, 46; ethics of, 97, 105, 108, 119, 205n6; v. facial-flap approaches, 107–8; facial manipulation, concerns over, 4, 90; fear of, 47, 49, 88; as fundamental disruption, 4; as gratuitous, 101, 103; health effects, 23; as hybridity, 4, 48, 125, 155, 165, 178; identity, concerns over, 6, 13–15, 24, 41, 47, 86, 90, 121, 165; immunosuppressant concerns, 94, 106–7, 109, 115, 119–20; as indeterminate state of being, 172; index, framework of, 2; as indulgence, 2; informed consent, 112–13; injured veterans, 121–22; insurance providers, conservativeness of, 147; interpersonal ethics, 86–87; as lifesaving, 18, 105; as makeover, 122–24; media coverage of, 97, 110, 117–18, 122,

133–45; mediatization of, 7–8; medically necessary v. elective, 2, 88, 93, 95, 103–4, 125; medicine, meaning of, 96, 116; military involvement in, 121, 142–43, 147, 150–53, 155; modernization, as metonym for, 54; and monstrosity, 48; morality of, 95, 100; neutralization, story of, 4; as never complete, 172; nonreversibility of, 2; objections to, 121; patients, on trial, 121; personal nature of, 2; psychological implications of, 41, 90–97, 104–5, 107, 109, 111, 113–14, 120–21; public opinion of, 89; quality of life issues, 96–97, 104, 109, 111, 114–15, 119–20; as radical disruption, 130; rejection, risk of, 91; risk-benefit analysis, 11, 47, 89–93, 106–7, 111, 114, 116, 120; self, meaning of, 9, 156; as self-actualization, 5, 125; as selfish, 2; sensationalizing of, 110, 116; suffering, ennobled by, 117, 121, 137; theories of face, 157–59; as therapeutic, 18; as third way, 14; and touch, 9, 164; trauma, normalization of, 6; visceral reaction to, 120; visibility of, 41–42; visual representation of, 8–9. *See also* makeover
facial allotransplantations (FAT). *See* face transplant surgery
faciality, 159
facial manipulation, 27–28; character manipulation, 16; horror films, 84–85; self, changing of, 23
Facial Surgery Research Foundation, 106
Farrell, Colleen, 15
fatness, 21
feminist studies, 20
Ferrari, Lolo, 171
flak, and clickbait, 96
Foucault, Michel, 29, 197n78
framing, journalistic practices, 96–97
France, 7, 53, 56, 82, 88, 103, 204n42, 206n26
Franju, George, 53, 56, 74, 202n10
Frankenheimer, John, 58, 202n23
Frankenstein (Shelley), 201n3

freak shows, 28, 30
Freund, Karl, 44–45

Garland-Thomson, Rosemarie, 16, 20–21, 169
gaze: cinematic gaze, and medical gaze, 45; male gaze, 171
Geiger, Jay, 205n11
gender, 5, 8
genetics, 33
Geroulanos, Stefanos, 53
Gillies, Harold Delf, 26, 33
Gilman, Sander, 15–17, 22–23, 40, 46–48
Gimlin, Debra, 193n12
Ghent (Belgium), 147
Goffman, Erving, 2, 20, 87, 163, 171
Gonder, Patrick, 43–44, 48
Goodley, Dan, 29, 193n14
Good Morning America (television show), 134–36, 138
GQ (magazine), 142–43
Grealy, Lucy, 213n32
Gréville, Edmond T., 44
Grindr, 218n27
Guattari, Félix, 2, 159, 168

Habermas, Jurgen: face, signification, iconic form of, 159–60; space in between, 178–79
Hallam, Clint, 102–3, 111, 206n26
Hamlet (Shakespeare), 1
Hands of a Stranger (film), 44
Hands of Orlac, The (film), 44
hand transplants, 32, 40, 89, 91, 94, 102–3, 109–12, 114–15, 141; "alien hand," 200n31; visibility of, 41–42
Haraway, Donna, 15, 29, 48, 165, 169–70
Hardison, Patrick, 148
Hartman, Rhonda Gay, 112–14
Harvard Ad Hoc Committee on Brain Death, 37
Hastings Center Studies, 37
Haupt, Clive, 41–42
heart transplants, 35–36, 148; and donors, 39

Hegel, Georg Wilhelm Friedrich, 167
Helfgot, Joseph, 133–34
Helfgot, Susan, 135–36
Her (app), 162
Herman, Edward S., 96
hermaphroditism, 172
Herndon, April, 21, 195n50
Herrick, Richard, 33–34
Herrick, Ronald, 33–34
Hersley, Chelsey, 30
Hettiaratchy, Shehan, 106–7
Hibbard, Kai, 129
Hippocratic Oath, 198n99
Hiroshima (Japan), 67
Hogle, Linda, 102
Homo Sacer: Sovereign Power and Bare Life (Agamben), 150
horror films, 56; facial manipulation, 84–85
Hours, The (film), 26
Humiliation (game), 1
Hunter, Mitch, 133, 139–40, 145, 148, 151
Hurley, Kelly, 43–44, 52
Hutchison, Iain, 106
Hutton, Laura, 198n84
hybridity, 5–6, 43, 48, 121, 125, 146, 155, 165, 175, 178; fear of unreliability, 47

identity, 53, 55, 62, 82, 131, 161, 179; changes in appearance, 13, 47; identity fraud, 162; identity transfer, 14, 32, 38–42, 45, 120–21; living dead, 75; public performances of, 163; and race, 51; as shifting, 47
immunology, 33
index: appearance, indexicality of, 179; of face, 3–4, 8, 12, 24, 50, 104, 160–62, 177, 180; face transplants, 2; indexical rupture, 162; as infinite, 162; vexing of, 4, 61, 87, 108, 161, 163, 165, 171, 176
India, 10–11
Institutional Review Boards (IRBs), 89–90, 103, 105–8, 111–12, 120, 204n1, 206n26
interrelationality, 85, 159
Introna, Lucas D., 163
Invisible Man, The (film), 55

Iraq, 151
IRL (in real life), 161–62, 179, 218n27
Italy, 32

Jackson, Cindy, 174
Jackson, Michael, 8, 171; boundaries, redrawing of, 173–76; as makeover monster, 175; in state of becoming, 173–75
Jaegger, Georgia May, 30
Japan, 7, 31, 53, 59, 82
Jaye, Lady, 172
Jehovah's Witnesses, 34
John and Kate Plus 8 (reality television show), 30
Jones, Meredith, 154, 171, 175, 194n28
Journal of American Medicine (journal), 37

Kael, Pauline, 56
Kass, Leon, 204–5n12
Kaur, Sandeep, 10–11, 17, 24, 28, 37–38, 191n1 (chap. 2), 192n4, 192n6
Kember, Sarah, 18
Kershaw, Abby Lee, 30
Kidman, Nicole, 26
kidney transplants, 33–35, 90–91, 148, 205n7; dialysis machines, 36–37
Kingsman: The Secret Service (film), 169–70
Kinja, 162
Kocher, Theodore, 33
Koswara, Dede, 125
Krajan, Ilija, 33
Kruvand, Marjorie, 95, 97

LaFrance, Marc, 96, 200n121
Lancet (journal), 40, 208n58
Lantieri, Laurent, 99
Lazar, Kay, 133–34
Lederer, Susan E., 32, 198n100
Lengelé, Benoît, 117–18
Lévinas, Emmanuel, 2, 68, 74, 85, 163, 176–80, 221n84; face, as abstract, 157; face, changeability of, 157; face, signification, iconic form of, 159–60; face, as surface of touch and exposure, 158; face, as two-

faced, 168; faceness, centrality of, 157; moment of encounter, 165–66; and other, encounter with, 157–59, 167, 172–73; and preknowledge, 166
Lewis, Tania, 125
Liesch, John, 194n28
Lindbergh, Charles, 33
Linker, Beth, 195n47
Little People Big World (reality television show), 30
Lock, Margaret, 31–32, 39
Lodge, David, 1
Longmore, Paul K., 52
Lowenstein, Adam, 53, 204n42

Mad Love (film), 44–45
Madonna, 169
Mains d'Orlac, Les (Renard), 44
makeover, 28, 130–31, 146, 153, 174; after portion, 128; big reveal, 5–6, 8, 16, 122, 124, 126; as earned, 124; face-transplant surgery, 122–24, 127; new self and identity, rebirth of, 126; process of, 25, 126; reality television, 122, 125–26, 154; self-improvement structure of, 128–29; transformation of, 125–26, 128–29; trauma of, 126. *See also* cosmetic surgery; face-transplant surgery; plastic surgery
Makeover, The: Reality Television and Reflexive Audiences (Sender), 146
makeover culture, 18
makeover television, 154; normative gendered bodies, as ideal outcome, 132; and women's culture, 132, 148–49
makeup, 27; as mask, 73; morality of, 203n39
Maki, James, 133–36, 138, 142, 148, 151, 213n32
Man without a Map (film), 203
Marbles, Jenna, 30
Martin, Barry, 152
Marvin, Carolyn, 15, 26, 150
masks, 29–30, 53, 55–62, 64, 70–72, 74–75, 78–80, 82–84, 100, 218n33; and makeup, 73
Mauss, Marcel, 34

McCarthyism, 59, 82, 202n23
McIndoe, Archibald, 33
McLaughlin, Janice, 19
McQuire, Scott, 131
Medawar, Peter, 33
Medicaid Bill HR 1, 36
medical documentaries, 127–28, 133, 149; conjoined twins, 131; normative gendered bodies, as ideal outcome, 132
Merrill, John, 33
mind, and body, 4, 9
Moebius Syndrome, 210n86
Monster (film), 26
monsters, 5, 45–46; fear of unreliability, 47; as threat, 43; as transgression, 43
Morrison, Wayne, 11–12
Moses, 166, 168; "hineini" moment of, 167
Moss, Barbara Robinette, 146
Mules, Warwick, 171
Murray, Joseph, 33
Myser, Catherine, 130–31

Nakamura, Lisa, 161
narrative binary, 12
Nash, Brianna, 141
Nash, Charla, 133, 141, 148, 151
Nash, Jamie, 138
National Association of Patients on Hemodialysis, 36
National Commission for the Protection of Human Subjects of Biomedical and Behavioral Research, 37
National Kidney Foundation, 36
Naugler, Diane, 17
New Wave movement, 202n10
Nip/Tuck (television series), 174, 194n28
Nixon, Richard M., 36–37
Norris, Richard Lee, 133, 141–44, 148, 151, 212n12
noselessness, 16, 25; prosthesis attachment, 26–27

Office of Naval Research, 151
Olney, Ian, 44

O'Neill, Robert, 43
online avatars, 156
online identity, 9, 162–63
Operation Smile, 24
Oprah Winfrey Show (television show), 136, 141
organ donation, 203n34; cross-racial and cross-sex, 41, 42
organ trafficking, 101
organ transplants, 13, 14, 25, 34, 38, 46, 92, 94, 102, 113–15, 117; and bioethics, 31–32; health access to, 37; identity change, 40–41; resistance to, 31–32
Orlacs Hände (film), 44
Orlan, 8, 171–73, 175–76
Orlan (body project), 172
other, 157–58, 163, 172–73; and agency, 180; and face, 159–60, 167, 177
Owen, Earl, 103

pandrogeny (positive androgyny), 172
Parameswaran, Radhika, 21–22
Patton, Cindy, 194n28
Paulette, K'aila, 198n100
Pearsell, Paul, 199n115
Peiss, Kathy, 16
Perry, Imani, 22, 169
Persona (film), 55
Peters, John Durham, 163–64, 217n21, 218n31
Petit, François, 109–10
PewDiePie, 30
physiogonomy, 177
Picture of Dorian Grey, The (Wilde), 47
Pinchevski, Amit, 163, 217n21
Plastic and Reconstructive Surgery (journal), 114
plastic surgery, 21–23, 25–26, 28, 37–38, 46–47, 171, 175–76, 193n12, 194n28, 201n142, 205n5; individuality, as celebrated, 29–30; necessary v. unnecessary, 13, 16–17, 24; performance of, 165. *See also* cosmetic surgery; makeover

Polan, Dana, 204n47, 204n48
Pomahač, Bohdan, 135, 138–39, 141, 144, 151–52
P-Orridge, Genesis Breyer, 8, 171–73
posthuman face, 158
Powell, Dilys, 56
Powell, Tia, 111
Powers, James, 58
Pratley, Gerald, 58
President's Commission for the Study of Ethical Problems in Medicine, 37
prosthetics, 26, 156, 161, 169–70
Psycho (film), 204n47

Queer Eye for the Straight Guy (reality television show), 212n6
Quigly, Isabel, 56

race, and identity, 51
racial profiling, 177
racism, 158, 168
Randolph, John, 202n23
Rashi, 219n45
Real Housewives franchise (reality television show), 171
reality television, 212n16; cosmetic surgery, 124; medical documentaries, 125. *See also individual shows*
reconstructive surgery, 24; as socially constructed, 19
Redon, Jean, 56
Reincarnation of Saint-Orlan, The (performance series), 171
relationality, 9, 87, 115, 164, 179–80
Renard, Maurice, 44
Rheingold, Howard, 161
Rivers, Joan, 8, 171; boundaries, redrawing of, 173–74; public persona of, 174; in state of becoming, 173–74
Rodgers, Herbert, 144
Rodriguez, Eduardo, 142–44, 151–52
Roth, Yoel, 161
Royal College of Surgeons, 111–12

Royal Free Hospital, 206n26
Ruck, Nora, 18

Sarwer, David, 23, 47
Sawyer, Diane, 135-38
Schöne, Georg, 33
Schweik, Susan M., 15
science fiction, 51
Scott, Matthew, 112
Seattle Center, 36
Seconds (film), 7, 49-50, 52, 56, 62, 64, 201n3, 202n14; deaths in, as punishment, 76; description of, 58-59; doubling, motif in, 82-83; and facelessness, 85; masking in, 82; medicine, dangers of in, 67-70; operation, institutional motif of in, 54; punishment motif in, 77-78; rebirth in, 59, 68-69; replacement, as story of, 54; space of exchange in, 63; surgery in, 63; transformation in, 61; transportation motif in, 53
Sekula, Allan, 15
self, 8, 46, 179; as changing, 6; face transplants, 9, 156; other, in relation to, 178; performativity of, 23, 28-29; self-improvement, 18, 31; self-manipulation, virtuous v. gratuitous, 14; self-modification, 21, 25; time, eradicating of, 76
Sender, Katherine, 18, 22, 132, 146, 148-49, 212n6
Sennett, Richard, 2, 23, 80, 87, 163
sexism, 158
Sharp, Lesley A., 39
Shattered (film), 55
She-Devil (film), 55
Shiel, Eileen, 118
Shildrick, Margrit, 131, 169
Shilling, Chris, 51
Sia, 218n33
Siebers, Tobin, 48
Siemionow, Maria, 105, 107-10, 138, 144
Sigan/Time (film), 55
Silverstone, Roger, 163

skin autographs, 32
Skin I Live In, The (film), 55
skin lightening, 22
Skintight: An Anatomy of Cosmetic Surgery (Jones), 175
Smart, Barry, 163
Smile Train, 24
Snyder, Sharon, 45-46
Sobchak, Vivian, 51
social death, 18, 179, 203n31
South Africa, 41-42
Sperati, Giorgio, 26
Stelarc, 173-74
Stolen Face (film), 55
Stone, Lara, 30
Stone, Sandy, 162
Stork, Travis, 144
Strunk, Jerry, 34
Sudnow, David, and social death, 203n31
Sushruta, 32
Swan, The (reality television show), 23, 38, 129, 176, 212n7
syphilis, and noselessness, 16, 25-26

Tagg, John, 131
Tagliacozzi, Gasparo, 32
Talley, Heather Laine, 18-19
Tarleton, Carmen, 133, 144-45, 148
tattooing, 27
Taylor-Alexander, Samuel, 91, 96-97, 206n20
teratology, 28
Teshigahara, Hiroshi, 59, 202n14, 203n25
Testalin, Sylvie, 89
Theron, Charlize, 26
Thomas, Abraham, 10
Thriller (video), 174-75
Thurman, Uma, 22, 196n57
Today (television show), 141
transcendence, 158, 166, 168
transplantation: as hybridity, 43; as miscegenation, 43; as monstrosity, 43; as transgression, 43

"Transplantation and Identity: A Dangerous Split?" (Carosella and Pradeu), 40
transplant films, 5, 7, 43, 48–50, 201n2, 201n9; body horror, as term, 44; and doubling, 51; existential crisis, 51; and identity, 51; as sci-fi horror canon, 51; transformation in, as zero-sum game, 52
transplant surgery, 47; consent, issue of, 34; development of, 32–38; and donors, 34–36, 39; gift giving, 34–35; identity change, 40; media representations of, 43; quality of life issues, 35; risk of, 35; self-harm, issue of, 34–36
Travolta, John, 1
Tuchman, Gaye, 96
Turkle, Sherry, 161
Turner, Bryan, 51
Turner, Graeme, 123

United Kingdom, 97, 106. *See also* Britain; England
United States, 7–8, 56, 69, 82, 91, 97, 103, 106, 133–35, 138, 147–49, 152–53
University of Maryland Medical Center, 142, 151–52
University of Maryland Shock Trauma Center, 142
"Until They Have Faces: The Ethics of Facial Allograft Transplantation" (Siemionow and Agich), 108–9

van Dijck, José, 29, 130–31
Vaughn, Matthew, 170
Von Furstenburg, Diane, 25

Washkansky, Louis, 35
Wegenstein, Bernadette, 18
Weiner, Annette, 34
Weins, Dallas, 133, 138–39, 141, 148, 163–64, 214n37
Weins, Scarlett, 138–39
What Not to Wear (reality television show), 129
White, Katherine, 39
Wiene, Robert, 44
Wilde, Oscar, 47
Wildenstein, Jocelyn, 174
Woman in the Dunes (film), 60
Woo, John, 1, 55
World of Warcraft, 162
World War I, 26, 33
World War II, 33, 51

Young, Nathan, 202n23

Zellweger, Renée, 22, 24, 176, 196n57, 221n80
Zirm, Eduard, 33
zombies, 204n45
Zylinska, Joanna, 18